GOTTLOB FREGE

On the Foundations of Geometry and Formal Theories
of Arithmetic

GOTTLOB FREGE

On the Foundations of Geometry and

Formal Theories of Arithmetic

Translated and with an Introduction

by Eike-Henner W. Kluge

New Haven and London, Yale University Press, 1971

Library of Congress catalog card number: 74-140533
International standard book number: 0-300-01393-0

Designed by John O. C. McCrillis,
set in Times Roman type,
and printed in the United States of America by
The Colonial Press, Clinton, Massachusetts.

Distributed in Great Britain, Europe, and Africa by
Yale University Press, Ltd., London; in Canada by
McGill-Queen's University Press, Montreal; in Mexico
by Centro Interamericano de Libros Académicos,
Mexico City; in Australasia by Australia and New
Zealand Book Co., Pty., Ltd., Artarmon, New South
Wales; in India by UBS Publishers' Distributors Pvt.,
Ltd., Delhi; in Japan by John Weatherhill, Inc.,
Tokyo.

Contents

Preface

The primary aim of this volume is to make available to the English-speaking reader Frege's more important writings on the notion of a purely formal theory and on independence-proofs for the axioms of axiomatic systems. With one partial exception —the 1903 articles from the series "On the Foundations of Geometry"—none of the selections included have previously appeared in English. This book, then, is intended to fill a lacuna in available Frege sources. But its purpose is more than merely bibliographical. Disregarding volume 2 of *The Basic Laws of Arithmetic,* of which to date no complete translation has appeared, nowhere but in the present selections does Frege give anything like a protracted and precise treatment of purely formal theories or of independence-proofs for axioms. This book, then, has the further purpose of letting Frege himself present aspects of his position that, although both interesting and important, nevertheless are generally unknown. In presenting his views, Frege also throws some light on issues he raises elsewhere: such issues as the relations between concepts and objects, predicative and nominative expressions, first- and second-level concepts, and first-level concepts and objects.

The make-up of this volume requires some explanation. First, a general point about the inclusion of papers and letters not written by Frege himself. This was prompted by the fact that except for Frege's letter to Liebmann and his article "On Formal

Theories of Arithmetic," all of Frege's writings included here
are parts of controversies between Frege and other scholars. It
was felt, therefore, that in the interest of a better understanding
of Frege's position and in fairness to all parties concerned, the
relevant writings of Frege's opponents should also be included.
Second, the particular choice of Frege's own writings. The
inclusion of "On the Foundations of Geometry" needs no justification: the series constitutes the core of this volume. All other
selections naturally group themselves around it. The Frege-
Hilbert correspondence is introductory to the series in both a
conceptual and an historical sense: it constitutes the genesis of
the "Foundation" series. In fact, the first part of that series
(dating from 1903) is but a more careful and explicit exposition
of what Frege argues in the correspondence. The Frege-Thomae
controversy was included because it illuminates some of the
points Frege makes in part 2 of the 1906 articles of "On the
Foundations of Geometry," where he replies to Korselt's for-
malistic defense of Hilbert. It should be read along with the
latter articles as well as with "On Formal Theories of Arithmetic," which is a separate article on the same topic. (Frege's
review of Husserl's *Philosophy of Arithmetic* is also relevant in
this connection.) As to Frege's letter to Liebmann, it was included for two reasons: First, because of its historical significance as Frege's own introduction to his correspondence with
Hilbert and hence indirectly to the "Foundations" series itself.
And second, because of its content: in it Frege gives an informal
exposition of the notion of a second-level concept and its relationship to first-level concepts and objects and pinpoints Hilbert's mistake in his "Foundations of Geometry" as resting on
an insufficient appreciation of the difference between first- and
second-level concepts. However, the importance of what he
says here extends far beyond the articles included in this volume.
The letter might profitably be read in conjunction with such
essays as "On Function and Concept" and "On Concept and
Object."

A word about the translation of *Bedeutung* by 'reference'. *Bedeutung* in German may mean either meaning, reference, or significance, depending on the context. It is systematically ambiguous. The present translation has the advantage of stressing quite clearly that for Frege, *Bedeutung* is a technical term whose core-meaning is that of 'reference'. However, that it is a technical term should not blind the reader to the fact that even in Frege's usage it retains certain associations. It is because of these that some of the peculiar difficulties raised by Thomae in his first paper arise. Any consistent rendition of *Bedeutung* by 'reference' would yield patent nonsense: one simply does not speak of the reference of a chess piece. Here 'significance' is the correct translation; and since part of its meaning is the same as that of 'reference', a consistent rendering by 'significance' initially seemed possible. Aside from consistency, it would also have had the advantage of permitting us to render into English, puns that hinge on the systematic ambiguity of *Bedeutung,* since 'significance' shares this ambiguity to a large degree. There are, then, substantive reasons for preferring 'significance' to 'reference'. However, there are two considerations that argued against such a translation. First, it would have been unfamiliar and thus might have caused confusion. Second, the plural of 'significance' is somewhat unusual and would have lent a peculiar ring to some passages. I have therefore used the more traditional 'reference' except in the article by Thomae.

The sources for the various selections are indicated at the beginning of each. The numbered footnotes in the text are Frege's, although their numbering was changed to run consecutively, beginning with number one in each selection. Asterisks indicate footnotes by the translator. The numbers set between slashes give the pagination of the German original, and cross-references in the text are to the pagination as it appears in the original.

I am indebted to my colleagues Karel Lambert and Peter Woodruff for their comments and suggestions on the Introduction and on the translation of "On the Foundations of Geometry," respectively. I should like to thank the University of California, Irvine, for a grant to type the final version of the manuscript, and I should also like to thank my wife, without whom these translations would never have been possible.

E.-H. W. K.

Irvine, California
July 1970

Introduction

Frege's series of articles entitled "On the Foundations of Geometry" constitutes the focal point of the selections included in this volume. The articles fall into two groups: one, dating from 1903, is a public presentation of the results of Frege's correspondence with Hilbert about the latter's Festschrift "On the Foundations of Geometry"; the other, dating from 1906, is a detailed reply to Korselt's defense of Hilbert's position against Frege's 1903 critique. The point of the polemic is this: In his Festschrift, Hilbert had attempted to construct an axiomatic system of geometry that would permit him to prove the independence and consistency of the axioms of Euclidean geometry while at the same time fulfilling "even the most stringent requirements of logic." Hilbert had constructed his system without the use of explicit, analytic definitions but instead had used so-called implicit definitions or definitions in use. Thus he had let his axioms introduce and delimit both the kinds of elements to be found in the system as well as the relations among them. The resulting axiomatic system was something that, as Hilbert himself remarked, could be regarded as a single, very complex definition. Hilbert then proceeded to give what he considered to be proofs of the independence and consistency of the axioms of his system. In particular, he attempted to prove that the Euclidean parallel-lines axiom is independent of the remaining axioms by showing that it can be replaced by other parallel-lines axioms without loss of consistency.

Frege's objections to Hilbert's enterprise fall into two parts: those directed against the nature and actual construction of the system itself—I shall call them *systematic objections;* and those directed against Hilbert's claim that given this system, he could prove the independence of the axioms by means of a substitution-maneuver—I shall call them *methodological objections.* Frege's discussion is complicated by the fact that Hilbert's defender, Korselt, suggested that Hilbert's system could be understood as an uninterpreted, purely formal system without content, which could be given different interpretations. This, Korselt claimed, would be consistent with Hilbert's intentions and would vitiate much of Frege's 1903 critique. (It is in connection with this notion of a purely formal system that the other two selections— the interchange with Thomae and the article "On Formal Theories of Arithmetic"—are relevant: they clarify some aspects of Frege's 1906 reply to Korselt.)

The aim of Frege's arguments in the selections included here is thus to make certain points on what I have called the systematic and the methodological levels, respectively. The relevance of these points, however, is not confined to their respective contexts, for in giving his analyses, Frege presents us with a detailed and precise discussion of such matters as the relationship between language and the world; the nature of definitions, explications, propositions, and axioms; the nature of concepts and objects, and their interconnection; and the relationship between first- and second-level concepts. Indeed, as very quickly becomes apparent, the very thrust of Frege's systematic and methodological objections hinges on the position he maintains on these matters.

Accordingly, I will begin my discussion with a brief exposition of the conceptual basis of Frege's critique, beginning with a quick sketch of his views on the relationship between language and the world, going on to consider what he takes to be the two fundamental types of linguistic units—what I call *nomina-tive* and *predicative* expressions—and then turning to Frege's theory of objects and concepts. Given these notions, I will then discuss Frege's views on definitions, explications, and axioms and give a brief exposition of his systematic objections to Hilbert's enterprise. In the second part of the Introduction I address myself to Frege's methodological critique of Hilbert's endeavor, sketching Frege's position on the notion of formal systems and indicating which notions he does and does not find acceptable. I will then present Frege's arguments that Hilbert's system may well be understood as a formal system in the acceptable sense, but that this does not alter the fact that it is a second-level, not a first-level, system; and that therefore the putative independence-proofs for the Euclidean axioms by means of the substitution-maneuver do not in fact prove the independence of these *first*-level axioms, but rather show something about the characteristics of the axioms of the *second*-level system.

Understanding Frege's position on the relationship between language and the world will be facilitated by introducing the notion of what Frege calls a perfect language.[1] Part of what he understands by this is quite familiar and needs little comment. It is encapsulated in the demand that all signs of such a language actually refer; that they succeed in naming.[2] But this aspect of the notion, familiar though it is, is subsidiary to the following: A perfect language is a language that reflects the logical structure of reality. This constitutes the real core of the notion of a perfect language, and once more Frege expresses it in the form of a demand: that every logical difference in reality be reflected in or by an analogous difference in the language itself. We find the clearest expression of this in Frege's more formal writings,

1. Cf. "On Sense and Reference" in Geach and Black, *Translations from the Philosophical Writings of Gottlob Frege* (Oxford, Blackwell, 1966), p. 70; see also pp. 69 ff., and p. 108 below.
2. "On Sense and Reference," pp. 70 ff., and passim.

such as the *Begriffsschrift*[3] and the *Basic Laws of Arithmetic;*[4] but it also plays an essential role in "On the Foundations of Geometry" below (see pp. 32-35, 60 ff.). In fact, it lies at the very heart of Frege's view of language *tout court*. For even when dealing with ordinary language Frege assumes that despite its logical and semantic imperfections, it contains a hard core that conforms to this demand. Hence the fact that although the demand is formulated as such with respect to a perfect language only, it is nevertheless extended to cover ordinary language as well—at least insofar as it is used in any kind of precise discourse. One might almost say that as far as Frege is concerned, ordinary language really is a perfect language corrupted by the intrusion of illogical and extralogical elements.

The signs or expressions of a perfect language fall into two exhaustive and mutually exclusive groups: those that are nominative and those that are predicative in nature. Examples of the former are proper names, definite descriptions, propositions, and so on; examples of the latter are predicative expressions in the ordinary sense, names of properties (inclusive of the copula),[5]

3. *Begriffsschrift, eine der arithmetischen nachgebildete Formelsprache des reinen Denkens* (Halle, a.s., L. Nerbert, 1879), p. v, where Frege indicates that it is an attempt to construct what Leibniz had called a calculus ratiocinator and a lingua characteria. Indeed, the very title of the book may be rendered as "Conceptually Perspicuous Notation." See also "Über die Begriffsschrift des Hernn Peano und meine Eigene," in *Berichte über die Verhandlungen der königlich-sächsischen Gesellschaft der Wissenschaften zu Leipzig, mathematisch-physische Klasse, 48* (1896), 361-78, esp. 362-65 and 370-71; and "Booles Rechnende Logik und die Begriffsschrift," Hermes, Kambartel, and Kaulbach, eds., *Nachgelassene Schriften* (Hamburg, Felix Meiner Verlag, 1969), pp. 9-52, esp. 9 ff.
4. *Grundgesetze der Arithmetik: Begriffsschriftlich Abgeleitet* (Jena, Pohle, 1893; all references are to the photomechanical reproduction, Hildesheim, G. Olms, 1962, 101 and esp. 136 ["For figures, arbitrary rules could be set up; in the case of signs, the rules follow from their references"], and passim.)
5. Frege regarded the copula as belonging to the predicative expression. Cf. *Nachgelassene Schriften*, p. 101; *Grundgesetze*, § 66.

and so on. Here it is important to preclude a possible misunderstanding: The division of expressions into these two groups is not on the basis of the grammatical roles they play in propositions. Thus, it is grammatically possible and indeed does happen that what has been called a predicative expression according to its nature occurs in what grammatically would be called a nominative or subjective role, without for all that losing its predicative character. *Mutatis mutandis* for nominative expressions. Rather, what this division reflects, and indeed is based on, is the fundamental difference in the natures of the entities they name.[6] That is to say, Frege locates the crux of the distinction between the two sorts of expressions in the fact that while nominative expressions are *saturated* or *complete,* thus being able to stand by themselves, predicative expressions are *unsaturated, incomplete,* or *in need of supplementation* (see below, pp. 33 ff.). However, Frege also holds that this difference in the nature of the expressions is not merely a linguistic phenomenon, but that it reflects and is reflected by an analogous distinction in the realm of references. What it reflects is the difference between what Frege calls an *object* and what he calls a *concept,*[7] where the difference between these two is absolute and primitive. Frege maintains that here no "schulgemässe Definition,"[8] no analytic definition—and hence, we shall see, no definition at all—is possible. However, as he says in his letter to Liebmann (see below), it is possible to "characterize" this difference: "The nature of

6. Frege called a referring sign a name. Cf. *Grundgesetze,* § 17: "I call *names* only those signs and sign complexes which are supposed to refer."
7. Strictly speaking, perhaps, I ought to have said *function* instead of *concept,* since a concept is but a special kind of function. However, since the distinction is not important here, and since a good discussion of it can be found in M. Furth's introduction to his translation of Frege's *Basic Laws of Arithmetic* (Berkeley and Los Angeles, University of California Press, 1964), pp. xxv-xxxv, I shall persist in this usage.
8. See "Funktion und Begriff," in G. Patzig, ed., *Gottlob Frege: Funktion, Begriff, Bedeutung: fünf logische Studien* (Göttingen, Vandenhoeck & Ruprecht, 1962, pp. 16-37), p. 28

xvi   Introduction

concepts can be characterized by the fact that they are said to have a predicative nature. An object can never be predicated of anything." Still differently, objects, like their linguistic counterparts (nominative expressions), are saturated or complete; whereas concepts, like their counterparts (predicative expressions), are unsaturated or incomplete. And just as, in the realm of expressions, a mere concatination of nominative expressions does not constitute a unit—a propositional whole—so in the realm of references, objects by themselves cannot enter into any sort of combination—The means of connection is lacking (cf. pp. 33-35 below). In the realm of expressions, this means is supplied by the predicative expressions, in that they are unsaturated or incomplete; in the realm of references, it is supplied by concepts that once more are characterized as incomplete or in need of supplementation.

The locutions *saturated* and *unsaturated, complete* and *incomplete* are, of course, metaphorical, and Frege attempts to spell them out. Yet he does so only in terms of another metaphor: concepts, as opposed to objects, are said to carry with them a *logical place* that requires being filled. Admittedly, this explanation does not clarify the notion of incompleteness very much, but it does bring out quite clearly one aspect of the nature of concepts that Frege means to emphasize: It is because concepts have such places, i.e. because they are unsaturated and incomplete, that they can carry the burden of ontological connection which Frege assigns to them. And it is because their linguistic counterparts are logically incomplete in an analogous sense, that they can be completed by other expressions, thus bringing about propositional wholes.

But this does not exhaust the role of logical places. They also provide the basis for grouping concepts into levels. Thus, some concepts carry with them places that can be filled only by objects, not concepts; others carry with them places that can be filled only by other concepts, not by objects (see p. 35 below). Those having logical places requiring completion by objects only are called first-level concepts or concepts of the first level; those concepts that can be saturated only by other concepts are called second-level concepts, and so on. Frege also maintains that the nature of these logical places is fixed and determined, so that objects can never substitute for concepts, and vice versa. We thus have an ontological hierarchy of concepts based on the nature of their respective logical places.

The concepts themselves may be distinguished according to their degree of internal complexity. That is to say, concepts at any level are either simple or complex; they may or may not have component concepts. The components of a concept Frege calls its *characteristics,* and he maintains that if the concept is complex, then it is constituted of its components concepts or characteristics.[9] To illustrate this point Frege adduces the example of stones making up a house. While not all concepts need be composite—in fact some are unanalyzable, simple, and primitive (cf. pp. 58-59, 60 ff., and passim below)—all objects are complex, at least to the extent that they have properties. The relationship between properties and concepts is such that whatever is a property of an object is a characteristic of the concept under which the object falls.

This introduces the next point: the relationship between concepts and objects. An object is said to *fall under* or *be subsumed under* a concept if and only if it has those properties that are characteristics of the concept.[10] If the object does have the properties—if it does fall under the concept—it is said to *complete* the concept. This relationship of *subsumption* holds only between first-level concepts and objects. Metaphysically and logically, an object cannot be subsumed under any other con-

9. Cf. *Nachgelassene Schriften,* pp. 247 and passim; pp. 4-5 below; pp. 35-36 below; *Die Grundlagen der Arithmetik: Eine logisch-mathematische Untersuchung über den Begriff der Zahl* (Breslau, W. Koebner, 1884), § 53; *Grundgesetze,* p. 3, etc.
10. Cf. pp. 4-5, 35 ff. below; cf. also *Nachgelassene Schriften,* p. 113, and "Concept and Object," in Geach and Black, pp. 50 ff.

cept. Linguistically, the result of completing a concept-expression of any level except the first by an object expression is not false but nonsensical (cf. p. 35 below). Neither can a concept be subsumed under another concept, no matter what the latter's level. To be sure there is a relationship between first- and second-level concepts that is analogous to that of subsumption, but it is *merely* analogous. It is expressed by saying that the first-level concept *falls within* the second-level concept. In fact, analogous sorts of relationships hold at all different levels of concepts. In each case, however, the relationship involved is unique. It is analogous to but logically distinct from the relationship of subsumption.

There is yet another relation, different from the above, which holds between concepts of the first level: that of subordination. The example Frege gives is that of the relationship between being a rectangle and being a square, as it is brought out in the proposition "All squares are rectangles." What he seems to have in mind is something similar to the relationship traditionally thought to hold between genus and species. Frege goes on to say that the characteristics of the superordinate concept (rectangle) are also characteristic of the subordinate one (square) (see p. 4 below).

Let us now turn from concepts to Frege's notion of objects. If the peculiarity of concepts is predicability and incompleteness, that of objects is impredictability and completeness. This is because Frege's objects are complete concepts; no more, no less. That is to say, Frege's objects are property-complexes on the Leibnizian model. However, since nowhere does Frege come out and say this in so many words, this must be argued. It can be argued in two ways: first, on the basis of his acceptance of a strong version of the principle of the identity of indiscernibles and his general and avowed Leibnizian tendencies; and second from his theory of the nature of truth-values and the references propositions.

The first argument goes as follows. Frege claims that it is

impossible for two or more objects to have identical properties and yet be distinct objects.[11] That is to say, he accepted the following version of the principles of the identity of indiscernibles:

$$(x) \ (y) \ (F) \ [(Fx \equiv Fy) \supset (x = y)]$$

To be sure, acceptance of this principle does not in and by itself entail that Frege's objects are no more than property-complexes; for the principle as such is quite compatible with the further postulate of substrata of properties, i.e. substances. However, two additional considerations make it extremely unlikely that Frege did advance such a postulate: (1) Tradionally, substances or substrata are postulated to fulfill either of two functions: to "support" properties, i.e. to be that which "has" properties; and to individuate, i.e. to account for the numerical diversity of objects. The first function, however, can be fulfilled by property-complexes of the Leibnizian sort without having to postulate substances. The second is already accounted for by the principle above. And as comes out quite clearly with his interchange with Thomae,[12] Frege does accept the principle in its individuating role. As he puts it: "If through this (abstraction of properties) the counting blocks become identical, then we now have only one counting block; counting will not get beyond 'one'." Clearly, Frege's point here holds if and only if the notion of an object is that of an entity constituted solely of properties. Otherwise, if an object did involve a nonqualitive entity such as a substance, or a substratum of some sort or other, the absolute qualitative identity of the objects would not make for their numerical identity, and counting would proceed beyond 'one'. It therefore follows that in Frege's systems a postulate of substances or substrata would not only add needlessly to the on-

---

11. Pp. 125 ff. below; *Grundlagen*, §§ 34-35 and passim; "Mr. Schubert's Numbers," in I. Angelelli, ed., *Gottlob Frege: Kleine Schriften* (Hildesheim, Georg Olms Verlagsbuchhandlung, 1967), pp. 245-47 and passim.
12. See below p. 125; cf. *Grundlagen*, §§ 33-35, § 45 and passim.

tology, it would also lead to a contradiction. (2) As was mentioned several times before, Frege is steeped in the Leibnizian tradition. In view of this, it seems only plausible to assume that on the point of substances he also follows Leibniz.

If the preceding series of considerations are mainly inclining, the following, second series of considerations carries demonstrative force. Frege claims that aside from fictional and similar contexts every proposition has a reference, and that the reference of a proposition is determined by (is constituted of) the references of its constitutive expressions (cf. pp. 32-34, 60 etc). Frege also claims that propositions can be classified into first-, second-, etc.-level propositions, where the precise level of the proposition depends on the particular level of the predicative expressions that plays the predicative grammatical role. Consider, now, a second-level proposition. Since a propositional whole comes about only through the completion of a predicative expression, and since, furthermore, a second-level predicative expression can be completed only by some other predicative expression, it follows that the constituents of this second-level proposition are names of (first- or second-level) concepts or properties. In other words, it follows from the above that the reference of the second-level proposition is constituted of the reference of names of (first- or second-level) properties. That is, it is a function solely of properties. Hence all second-level propositions[13] have property-complexes as their references. However, Frege also claims that propositions are names of objects, i.e. their references are objects. Nor do second-level propositions constitute an exception to this. Therefore, at least in the case of the references of second-level propositions, objects are identified as property-complexes. The preceding result can be generalized for all propositions irrespective of level. For no matter what the level, the objects to which propositions refer are truth-values:

13. In these and similar contexts, the proviso about fictional contexts, etc., indicated at the beginning of this paragraph, will be assumed as understood.

the True and the False. Each proposition refers to one of these; nor does Frege admit more than one True or more than one False.[14] That is to say, all true propositions, no matter what their level, refer to one and the same entity: the True. The same holds, *mutatis mutandis,* for all false propositions. But remember that the objects referred to by second-level propositions were identified as property-complexes; and that therefore, seeing that the objects referred to are truth-values, the truth-values referred to by second-level propositions have been identified as property-complexes. Given what was just said, that there are only two truth-values, it follows that both truth-values are property-complexes. Consequently, the objects referred to the propositions in general are property-complexes.

However, this analysis of objects as property-complexes holds not merely for truth-values but for any object whatever. This can be seen from the following. Frege states that the distinguishing characteristic of objects as opposed to concepts is that the former are complete and therefore can complete first-level concepts. In fact, Frege insists that any object whatever can complete (be an argument of) a first-level concept.[15] Now assume that Frege did admit objects of the substratum-cum-property type. Clearly, these would be distinct in nature from the objects he calls truth-values, which, as was just shown, are property-complexes. However, being objects, these new objects could be arguments of first-level concepts—their nature as objects guarantees this. But if they were such arguments, then the result of completing a first-level concept by such an object would be a truth-value whose nature would be radically distinct from that resulting from the completion of the concepts by truth-values of the traditional kind. Yet supposedly, there are only two truth-values! Frege would therefore be faced with a contradiction. The only way out would be for him to postulate distinct sets of truth-values; in fact, to postulate a whole hier-

14. Cf. "Sense and Reference," pp. 63 ff.
15. "Funktion und Begriff," p. 27.

archy of truth-values commensurate with all the possible varia-
tions on the completion of first-level concepts. No indication
whatever of anything even remotely like this is to be found
anywhere in the writings of Frege; quite the contrary.[16] Therefore it follows that in order to preserve Frege's doctrine of
truth-values, all of his objects must have the same nature: they
must be property-complexes.

The diagram in Figure 1 codifies Frege's position on the
hierarchical relationships between objects, first-level concepts,
and second-level concepts, etc.

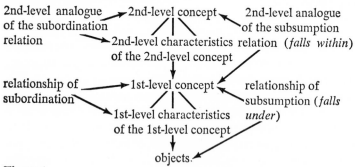

Figure 1

An analogous situation obtains in the realm of expressions.

With this background, let us now turn to Frege's systematic
and methodological objections to Hilbert's enterprise, beginning
with the systematic critique. Recall that in the construction of
his system Hilbert had not used explicit, analytic definitions
but instead implicit definitions or definitions in use. That is to
say, he had let his axioms introduce new terms as well as delimit the kinds of relations that can hold between them. Frege
objects to this. He distinguishes sharply between explicit definitions, or what he calls definitions *tout court*; implicit definitions,
or what he calls *explications;*[17] and axioms. Definitions are stip-

16. *Grundgesetze,* § 5; "Funktion und Begriff," pp. 27-31.
17. Frege uses this term, as distinct from 'explanations'.

ulations of the references of signs that so far have no reference within the system. They are given within the system. Explications, on the other hand, are not given within the system but are presupposed by it. Axioms, finally, are basic, unprovable statements within a system that express basic facts of our intuition.

Frege's position on definitions may be described thus: Assume an axiomatic system having a basic stock of referring expressions of both the nominative and the predicative type. Complex expressions can be formed from these in accordance with the appropriate rules, so that the references of the complex expressions are determined by the references of their constitutive expressions. However, such complex expressions frequently become too unwieldly, taking too long to utter or write; so a new sign is introduced by means of a proposition which stipulates that it be another name of the entity referred to by the previous complex expressions. The new sign can now occur in place of the complex expressions, and the proposition in which it is introduced is called a definition (cf. pp. 23 ff., 60, below). Considered as introducing a sign, a definition is neither true nor false; rather it is acceptable or not acceptable. It is no more and no less than the stipulation of the reference of a sign that, so far as the system is concerned, hitherto has had no reference. Once it is accepted, however, it is a proposition that is true of itself and that can be used in an inference as can a principle or axiom. The epistemic value of such a proposition is of course limited. As Frege says, no definition increases our knowledge, for once accepted, it amounts to no more than an example of the law of identity: $a = a$ (see pp. 24 ff. below). But whereas a definition has very limited epistemic value, it does have pragmatic value: it makes for greater ease of exposition and construction. However, this is merely peripheral.

> The real importance of a definition lies in its logical construction out of primitive elements. . . . The insight it

permits into the logical structure is not only valuable in itself, but also is a condition for insight into the logical linkage of truths (p. 61 below).

There are several things Frege requires of definitions. First, definitions must not mix levels, that is, no definition may include, as characteristic-expressions in its definiens, concept-expressions of distinct levels (see pp. 32, 35 ff., esp. 36, below). The reason for this is simple enough. What is defined are concept-expressions, and the references of the definienda are determined by the references of the characteristic-expressions contained in the respective definiens. Now at the first level, the relationship between references of the characteristic-expressions that are part of the definiens and the reference of the definiens as a whole can be expressed by saying that the concepts referred to by the former are subordinate to those referred to by the latter. However, this relationship can hold only among first-level concepts, and in the way indicated above. An analogous situation obtains at all other levels (see pp. 35-36 below). Since concept-expressions must reflect this fact, no definition may mix, as characteristic-expressions, concept-expressions of different levels. Frege gives a famous example where this requirement is not met: the ontological argument for the existence of God (see pp. 18 ff. and 32 ff. below). Here existence—a second-level property or concept—is mentioned in the definition of God as a characteristic of the first-level concept of an omniscient, omnipotent, omnipresent, etc. being. Such a definition is ill-formed, and the concept to which it allegedly refers is a metaphysical impossibility.

The second requirement is that the definition of a sign must determine the reference of that sign completely and unambiguously.[18] Unambiguously, because otherwise the sign might have a vacillating usage, possibly resulting in sophisms. Completely, for two reasons. First, because the reference of a defined sign

18. Cf. pp. 61-69 below; *Grundgesetze,* §§ 56 ff.; etc.

is a (complex) concept.[19] Therefore if the definition did not determine the logical structure of its definiendum completely, it would not determine that of its reference either, and hence we should have a concept where it would not be determinate for any object whatever, whether it falls under that concept or not. This would constitute a breach of the law of the excluded middle, which cannot be allowed. Second, and functionally related to the preceding, such an incomplete determination would mean that the reference of the definiendum would be metaphysically incomplete. But not incomplete in the way in which a concept as opposed to an object is said to be incomplete; i.e. it would not be incomplete in the sense of being unsaturated. Instead, it would mean that even qua concept, it would be a metaphysically incomplete entity—which is impossible (see pp. 68, 73 below). He expresses both of these points in volume 2, § 56 of his *Basic Laws of Arithmetic* as follows:

> The definition of a concept (possible predicate) must be complete; it must determine unambiguously for any object whatever, whether it falls under the concept (whether the predicate can truly be asserted of it) or not. Thus there must not be any object concerning which, given the definition, it remains in doubt whether it falls under this concept, even though it may not be possible for us humans with our deficient knowledge to decide the question. We can express it metaphorically like this: A concept must be sharply delimited. If we let an area in a plain stand for the extension of a concept, then although this analogy must be used with caution, it nevertheless serves our present purposes well. To a concept which is not sharply delimited there would correspond an area which did not everywhere have sharp boundary-lines but in places merged vaguely into the surrounding area. This would not really be an area

19. Cf. what was said about characteristics and characteristic-expressions in definitions; see also below.

at all; and similarly, an imprecisely defined concept would
be called a concept unjustifiedly. Logic cannot admit as
concepts such concept-like constructions: It is impossible
to lay down precise laws for them. The law of the excluded
middle is really no more than a different formulation of
the demand that the concepts be sharply delimited. Any
object Δ whatever either falls under the concept φ or it
does not: *tertium non datur.*

Nor does this requirement entail that, for Frege, the reference of the defined expression actually be exemplified; i.e.,
that here and now there must be an object that has the properties referred to by the characteristic-expressions constitutive
of the definiens and that hence falls under the concept in
question. As Frege puts it, the concept may very well be empty:
there may in fact be no object that falls under it.[20] An example
would be the concept of a man 50,000 feet tall. Still, the concept in question would be—would have to be—no less determinate, nor would its expression be any less ambiguous and
definite, because of this.

Closely connected with the preceding—indeed, part and
parcel of it—is Frege's position on what he calls piecemeal
defining.[21] The latter occurs whenever an expression is defined
for a particular case or series of cases only and later is further
defined to cover other cases or series of cases. Frege rejects
this method of defining for three reasons. The first is pragmatic:
if terms are defined in this way, then there will be no guarantee
that contradictions among the various partial definitions will
be avoided. That is, if contradictions are avoided, this will
be a matter of fortuitous circumstance only, for "this method
provides no basic guarantee that they will be ruled out."[22]
Second, partial definitions of a term would (trivially) define

20. Cf. *Grundgesetze,* pp. xiv, 3, etc.
21. Ibid., §§ 60 ff.
22. Ibid., § 57.

it only ambiguously and indeterminately—the concept would not be sharply delimited. Consequently, as we saw above, we should have a breach of the law of the excluded middle. For until there is a complete definition, it will not be determinate for any object whatever, whether it falls under the concept in question or not. Third, if this method of defining were allowed, the reference of a definiendum would change with each addition to the stock of its definitions. But then the proofs and theorems constructed using the term as only partially defined will not hold for the term as defined at the end of the piecemeal defining procedure. What we thought we had proved, we would not have proved at all.

A fact that emerges quite clearly from the preceding is that defining involves construction out of more primitive expressions. Obviously such a process of construction cannot go on indefinitely; definition must come to an end somewhere. Korselt makes much of this in his critique of Frege, but Frege is only too happy to admit the point. In fact, he claims that we must admit logically primitive expressions that are unanalyzable and hence cannot be defined.[23] In their case, the best we can do is given an *elucidation* or *explication;* for a definition, being a function of more primitive elements, would contradict the very primitiveness assumed for these expressions. Unlike definitions, explications cannot be given within a system but instead are presupposed by it. They belong to the preamble—to the propaedeutic, as Frege calls it. Of course, even explications presuppose the intelligibility of at least some of the terms in-

23. Cf. pp. 59-60 below. Frege uses the phrase "logically primitive element" for both indefinable expressions and their references. However, at bottom he does distinguish between the two (cf. pp. 59 and 60 below). In the light of what he says about the characteristics of concepts (cf. pp. 35 ff.), this ambivalent usage seems to be a rather imprecise reflection of the thesis that a perfect language mirrors the world. But no matter how the phrase is understood, what it refers to *is* logically primitive and *is* indefinable.

volved in giving them; hence not even here can we escape the necessity of simply presupposing some terms as known (see pp. 57 ff. below, esp. 59-64). Perhaps ostension enters; we just don't know—Frege never talks about this. But it is unimportant here, for the distinction between definitions and explications stands even without this.

Just as Frege stresses the difference between definitions and explications, so he emphasizes the difference between definitions and axioms.[24] He accepts the traditional notion of axioms as thoughts (propositions) that express basic facts of our intuition and are therefore true without being provable by a chain of logical inferences. Axioms, then, as opposed to definitions, do extend our knowledge. And they do not stipulate the references of any of the terms occurring in their expressions, but rather presuppose that all the terms occurring in them are known. Or, what amounts to the same thing for Frege, all terms occurring in the statement of an axiom must either be primitive within the system or have been introduced by definition prior to the statement of the axiom. Thus axioms do not introduce new terms. But there is still another difference between axioms and definitions. While we must make certain that definitions *will not* contradict one another or lead to later contradictions, axioms, being expressions of basic facts of our intuition and therefore true, *cannot* contradict one another. It is clear from all this that axioms occupy a rather special place in Frege's scheme of things. They constitute the ultimate ground of all inference within a given system.

We can now appreciate Frege's systematic critique of Hilbert's enterprise. Frege objects to Hilbert's rather fast and loose construal of axioms, definitions, and explications. Hilbert let his axioms introduce new terms; to Frege, this is anathema. His point here is not so much that this is not in accordance with

---

24. It goes without saying that Frege considers axioms to be distinct from explications.

the traditional meaning of these terms—an objection that could easily be met by a change in terminology—but that the sorts of propositions involved have distinct functions and natures, and that Hilbert conflates these. Both Hilbert and Korselt maintain, or at least accept, that over and above expressing basic unprovable facts of our intuition and therefore serving as basic propositions of a system, axioms also introduce new terms. Very well, argues Frege; let us consider axioms as also introducing new terms! There are only two ways in which new terms can be introduced: by definition or by explication. If axioms explicated rather than defined the terms they allegedly introduce, it would follow that they could neither be parts of the system nor be used as the bases of inferences in the way in which Hilbert's axioms are supposed to function: explications of the terms of a system do not belong to the system itself, but to its propaedeutic. On this interpretation of the function of axioms, then, Hilbert's axiomatization would collapse. However, it is fairly clear that Hilbert did not intend them as explications, but as definitions. Assume, then, that axioms define. In that case, there are two possibilities: Either each axiom defines the term(s) it introduces completely; or it gives a piecemeal definition, and the term(s) thus introduced is (are) completely determined only given the totality of axioms.[25] The first can be ruled out, since it conflicts with Hilbert's actual practice as well as with his explicit claims (cf. pp. 11 ff. below, esp. 13). The second, however, has unfortunate consequences. First, axioms would no longer be informative and increase our knowledge, but would be mere examples of the law of identity: $a = a$. Second, they would be true not because they express basic facts of our intuition but because they are definitions. They would be definitionally, rather than intuitionally, true. Also, on the present hypothesis, the reference of a new term

25. The reason for this being that the determination of a term is a function of the other terms occurring in the axiom, and these will be completely determined only by the totality of axioms.

will be functionally related to the references of the other terms occurring in the axiom; whence it follows that the references of the terms defined by the axioms would change with each addition of a new axiom or the substitution of a new axiom for an old one. Consequently, what is proved using a term introduced in one, say the first, axiom would not be proved about the reference of that term as it is finally defined by the totality of axioms. Only by an equivocation could the initial statement of the proof be thought to be about the same entity referred to by the term as finally defined. From this it follows further that Hilbert's independence-proofs for the axioms fail. For these proofs require that some axioms be assumed valid and others invalid; that some axioms, e.g. the Euclidean parallel-lines axiom, be replaced by different axioms, while others remain the same. Since, *ex hypothesi,* the references of the terms involved in the expressions of the axioms are functions of the totality of axioms, this proof-procedure amounts to no more than the retention of some axiom-*expressions,* whereas the axioms themselves are changed by the substitution of the new axiom. A new and distinct set of axioms results from such a substitution; therefore the axioms involved in the actual proof will not be the axioms whose independence is supposed to be proved.

Leaving this systematic critique, aimed at the actual construction of Hilbert's system, let us now turn to Frege's methodological critique, directed against the very method employed by Hilbert in his enterprise. Hilbert had intended to prove that the Euclidean parallel-lines axiom is independent of the remaining axioms. To this end he had constructed an axiomatic system with implicit definitions; more precisely, a system where the axioms do not merely impose conditions both on the kinds of elements to be found in the system as well as on the relations holding between them, but where these axioms also introduce the very elements of the system itself. He had

then shown that the substitution of other parallel-lines axioms
in place of the Euclidean one did not result in a loss of consistency. This, he concluded, established what he had set out
to prove.

In his 1903 series "On the Foundations of Geometry,"
Frege raises the following objection to this procedure. The
characteristics of the concepts defined by Hilbert's axioms are
not of the first but of the second level (see pp. 19, 36 ff.,
below); therefore what these axioms actually define[26] are
second-level concepts and relations, and what Hilbert has actually constructed is a second-level system. The concepts of a
first-level geometry, e.g. of Euclidean geometry, will fall within
those defined in this second-level one, and the particular
geometry itself will be no more than a special case of the latter.
In fact, as Frege says, "Euclidean geometry presents itself as
a special case of a more inclusive system which allows for
innumerable other special cases—innumerable other geometries,
if that word is still admissible" (see p. 37 below). Now, since
for Hilbert the axioms of a system define and delimit its concepts and relations, the substitution of one axiom for another
in a given system will result in different concepts and relations
being defined; in short, in a different system. Applied to the
case of Euclidean geometry, this means that the substitution of
a different axiom for the parallel-lines axiom results in a different first-level geometry. Hilbert had thought that the fact that
this sort of substitution does not result in inconsistency shows
the independence of the Euclidean parallel-lines axiom. Frege
argues that given the above we can see that it shows no such
thing; what it does show is the independence of the characteristics of the second-level concepts defined by the axioms of
the general system of which Euclidean and non-Euclidean

26. For the moment we disregard, with Frege, any difficulties about
defining axioms.

geometries are but special cases. In fact, Frege argues that given Hilbert's method of constructing axiomatic systems (by means of implicit definitions) and the very procedure employed, the independence of Euclidean (first-level) axioms *cannot* be proved; the impossibility lies in the very nature of the case. This point emerges even more clearly in Frege's 1906 reply to Korselt's defense of Hilbert's enterprise.

Korselt had argued (see text below) that Hilbert's notion of definitions and axioms was quite unexceptionable and that the system he had constructed would serve the purposes for which it was designed if only it were understood as intended: as a purely formal system which could be variously interpreted. As such, various particular first-level geometries differing solely in their parallel-lines axioms would result from it. And this fact, he argued, did show what Hilbert thought it did: the independence of these axioms.

Frege, after restating his previous systematic critique, nevertheless proposes to take Korselt's suggestion of a purely formal system seriously and begins by distinguishing between two notions associated with the phrase 'purely formal system'. The first notion, unacceptable to Frege,[27] is that of a system of merely physical entities, manipulated according to certain arbitrarily stipulated rules. Frege argues that understood in this way, a purely formal system would not be a meaningful one in any ordinary sense of the phrase, for as such it would have to contain meaningful signs, not merely physical entities, as well as propositions that were not mere concatenations of meaningless signs. In fact, understood in this sense, not even logical operations could be involved in such a system, for like propositions and meaningful signs, they require the presence of thought or sense. The latter, however, cannot be present consistent with the assumption of the uninterpreted and purely

27. Here the Frege–Thomae interchange and the article "On Formal Theories of Arithmetic" are relevant (see below).

formal nature of the system. Furthermore, strictly speaking even the claim that there are arbitrary rules within such a system is mistaken, for they too require that sense be introduced and that the system be given a certain minimal interpretation. To be sure, continues Frege, such a system may give the appearance of being more than a mere complex of physical entities. However, this is appearance only, brought about by the fact that the signs used in the system are already familiar to us from other contexts where they do have a sense, thus fostering the illusion that they are meaningful even here. This appearance will be dispelled the moment we substitute hitherto unfamiliar signs for the familiar ones; signs like these:

⊢   §   ∪   £   ℈℃   >   <   Χ   ∩   ∪ $ ∩   ⊨

Clearly, concludes Frege, the suggestion that Hilbert's system be construed as a purely formal system cannot be meant in this sense but must be understood differently. And this brings us to the second sense of the phrase, which is acceptable to Frege. His exposition of it involves a distinction between propositions and pseudo-propositions on the basis of sense and reference. Let us consider this briefly.

Frege distinguishes between the cognitive content of a proposition—its sense—and that to which it refers. Reference was discussed before and needs no further comment. By sense Frege means not something subjective and unique to a particular thinker, but rather something objective which can be intellectually apprehended by many. It is through its sense that a proposition refers to the world, just as any expression whatever refers through its sense. The diagram in Figure 2, taken from a letter from Frege to Husserl,[28] illustrates this relationship:

28. May 24, 1891; I thank the editors of the *Nachgelassene Schriften* and Frege's *Wissenschaftlicher Briefwechsel* for making a typescript copy of the letter available to me. The letter itself is to appear in Hermes, Kambartel, Kaulbach, eds., *G. Frege: Wissenschaftlicher Briefwechsel*.

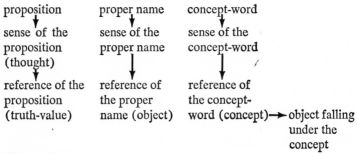

Figure 2

Now consider what are generally called propositions. Frege claims that they can be divided into two kinds: those that are propositions in a grammatical sense only, and those that are propositions properly speaking. Frege's distinction is based on the fact that something may be a proposition from a purely grammatical standpoint without having either sense or reference when considered in itself. To have both, however, is the mark of a proposition properly so called—what Frege calls a *real proposition.* He calls merely grammatical propositions *pseudo-propositions* and states that they are most frequently encountered as antecedents and consequents of conditional propositional complexes. What he has in mind can be illustrated by the following complexes:

1. If something is greater than 1, then it is a positive number.
2. Anything smaller than 1 is a negative number or zero.[29]
3. If a^2 is a whole number, then $(a \times (a - 1))$ is an even number.

Although each of these, considered as a whole, is a real proposition, each may be decomposed into propositional parts. Abstracting from the logical expressions involved—'if', 'then', etc.—each of these parts will be a pseudo-proposition by virtue of containing dummy expressions or letters, i.e. letters or expres-

29. Frege claims that the real structure of 2. is conditional: If something is smaller than 1, then it is a negative number or zero.

sions that do not actually refer. It is because they contain the latter that no pseudo-proposition (considered in itself) has either sense or reference. It is only the complexes constituted of pseudo-propositions that have both sense and reference; and the reference it has is in each case one object—a truth-value—just as the sense expressed is in each case only a single thought.[30] Frequently it happens that the fact that something is a pseudo-proposition is obscured by grammatical appearances. The following is Frege's example:

Let a be a whole number. $(a \times (a-1))$ is an even number.

What we have here are actually two pseudo-propositions which, when taken together, form a conditional complex having a complete reference and sense, whereas alone neither pseudo-proposition does. The real nature and structure of their relationship can be expressed more precisely and perspicuously like this:

If a is a whole number, then $(a \times (a-1))$ is an even number. Frege calls the pseudo-proposition occurring as the antecedent of such a complex, the *antecedent pseudo-proposition;* he calls the one occurring as its consequent the *consequent pseudo-proposition.*

Although pseudo-propositions like the above have neither sense nor reference, still they do contain referring parts that have both. These parts are those that remain after the removal of dummy expressions—'something', 'anything', 'it', etc.—or letters—'a', etc.—from the pseudo-propositions. So, for example, the referring parts of the antecedent pseudo-propositions of the examples above would be,

. . . is greater than 1.

30. Cf. p. 71 below. It may seem as though—to take but one example—"something is greater than 1" does express a thought; namely, that there is something which is greater than 1. But as Frege says, "it is not in this sense that the grammatical proposition occurs as antecedent of the propositional complex." (pp. 69 ff. below)

. . . is smaller than 1.
. . . is a whole number.

Clearly, these are unsaturated expressions; hence their references, too, will be unsaturated. That is, they will be concepts. And whereas the function of the concept-expressions contained in these pseudo-propositions is to refer, such is not the function of the letters and dummy expressions. Their function is "to lend generality of content to the whole proposition" of which the pseudo-propositions are parts (p. 71 below). The thought expressed by the resulting propositional complex is a general one about the relationship between the concepts referred to by the referring parts of the antecedent and consequent pseudo-propositions respectively. In the case of 1. above, this relationship is that of the subordination of the concept *greater than 1* under the concept *positive number*. An analogous relationship holds for all conditional complexes of like structure.[31] "And thus it comes about that the whole propositional complex expresses a true thought, even though it contains a letter that signifies nothing" (p. 71 below). However, although neither antecedent nor consequent pseudo-propositions express a thought by themselves nor, therefore, have a reference, they may nevertheless be turned into real propositions that do. This may be effected by replacing the dummy letters or expressions that merely indicate with actually referring expressions.[32] Thus in the case of 1., if we substitute the name of a number (of an *object*) uniformly throughout, we obtain a complex conditional proposition composed of real propositions in place of the old pseudo-propositions. These real propositions will be about particular entities. Appropriately enough, Frege calls the procedure involved in deriving these propositions an *inference from the general to the particular* (see pp. 74, 75, etc., below).

31. The exact relationship involved will depend on the nature and level of the concepts involved.
32. Of a kind determined by the nature and level of the concept-expression contained in the respective pseudo-propositions.

Similarly in all other cases. Of course, neither antecedent nor consequent real propositions thus derived are themselves asserted: only the whole conditional complex of which they are parts is. However, given the general laws of logic, if the resulting antecedent propositions are true, they may be dropped, thereby resulting in "a particular proposition containing fewer antecedent propositions" (p. 75 below).

Frege then defines a *formal theory* or a *purely formal system* as a system of general theorems that agree in their antecedent pseudo-propositions. Given standard logical techniques, we can derive from these one complex proposition consisting of the antecedent pseudo-propositions of the theorems and one—generally complex—consequent pseudo-proposition. Such a proposition—such a theory—will be general and as such will concern relationships among concepts. However, it can be particularized in the way indicated above—by the simple expedient of substituting referring expressions uniformly for the dummy letters or expressions. Since various distinct substitutions are possible, we can derive distinct, particular, lower-level theories from the general one. This fact may engender the illusion that we have an ambiguity at the higher level and that what Frege calls an inference from the general to the particular is no more than a clarification of this ambiguity. Actually, however, this multiplicity of possible lower-level theories is a *logical* phenomenon which depends on the fact that more than one first-level concept can fall within a second-level concept, and analogously at all other levels. As Frege emphasizes time and again, any appearance of ambiguity, any appearance that ambiguity is necessary, results from a failure to appreciate the distinction between first- and second-level concepts, etc.[33]

33. Korselt talks about various *interpretations* of the higher-level theory in order to arrive at distinct lower-level ones, and claims that before the higher-level theory is thus interpreted, it is ambiguous. Frege argues that this appearance of ambiguity is in no small measure a result of the use of 'interpretation'; for as ordinarily understood, interpretation—

Is Hilbert's system then a purely formal system in the above sense? Frege argues that it may very well be understood as such. Hilbert's so-called defining axioms, which on any other interpretation run afoul of Frege's systematic critique, on this interpretation turn out to be neither definitions nor axioms, but instead the antecedent pseudo-propositions of the propositional complex that is the whole system. As pseudo-propositions, the fact that they contain otherwise undefined meaningless expressions—'point', 'plane', 'straight line', 'lies in', 'lies on', etc. (cf. pp. 80-81, etc., below)—occasions no difficulties; on the contrary, it is a necessary condition of their status as antecedent pseudo-propositions. Of course, given that the so-called axioms are of this nature, it follows that what Hilbert had taken to be theorems of his system are not really theorems either, since (1) they too contain the dummy expressions mentioned above; and (2) the axioms from which they are allegedly derived are pseudo-axioms—pseudo-propositions that are neither true nor false—for which reason no inferences can be based on them (cf. p. 86 below). The system Hilbert had constructed can therefore be viewed as a general, second-level, purely formal system where the alleged axioms and theorems are antecedent and consequent pseudo-propositions respectively, and where the alleged inferences are merely schemata of inferences that result from the substitution of actually referring expressions for the dummy expressions contained in the system (see pp. 82 ff. and passim below).

This introduces the next point. Understood as a formal theory in Frege's sense, Hilbert's system can be variously instantiated. That is to say, by means of various inferences from the general to the particular, the pseudo-axioms functioning as antecedent pseudo-propositions can become distinct real axioms of differ-

---

Deutung—is indeed the removal of an ambiguity. This explains Frege's preference for the locution 'inference from the general to the particular'; which, by the way, he also takes to be more expressive of what actually occurs in these cases.

ent first-level theories; the pseudo-theorems functioning as consequent pseudo-propositions can become different real theorems of those first-level theories; the inference-schemata of the general theory can be turned into real inferences of the particular theories; and so on. Of course it remains a separate and thus far unsettled question, whether any particular first-level theory will be a true description of reality. That questions cannot be settled by proving the general theory of which these first-level theories are particular cases. For the former is true and is proved quite independently of the truth of any particular axioms of any first-level theory that may be inferred from it (see p. 92 and passim below).

Not only can Hilbert's system, understood in this way, be thus instantiated, but more particularly, by means of inferences from the general to the particular, various and distinct first-level geometries can be arrived at which differ in their respective parallel-lines axioms. Now both Hilbert and Korselt were of the opinion that in showing that the substitution of different parallel-lines axioms for the Euclidean one results in different but nevertheless consistent particular geometries, they had shown something about Euclidean geometry itself: namely, that the parallel-lines axiom is independent of the other axioms. Frege, however, quite correctly points out that given the nature of the case they had shown no such thing. All they had thereby shown is that various consistent first-level theories can fall within the second-level theory Hilbert had constructed. To be sure, Frege continues, to have shown this is important. But far from establishing the independence of the Euclidean—first-level—axioms, this proof is not even concerned with these but rather with the characteristics of the concepts of the relevant pseudo-axioms from which the particular axioms are inferred. And what it shows about these characteristics is that *they* are independent. A separate and quite different sort of independence-proof will be needed to establish the independence of the first-level axioms. For the procedure employed by Hilbert, depending as it does on the substitution-maneuver—on the possibility of distinct inferences from the general to the particular, as above—could not possibly be employed in the case of anything except a purely formal system of the present kind.

Frege's methodological objection can be summed up in another way. The procedure employed by Hilbert centers about a substitution-maneuver, which, if it is to go through, requires that Hilbert's system be understood as a purely formal system in the way indicated by Frege: as a second-level theory. Here, and here only, is the procedure adopted by Hilbert appropriate, since that procedure amounts to no more than showing that distinct inferences from the general to the particular do not lead to contradiction. But showing this shows nothing about the first-level axioms whose independence Hilbert set out to prove. What it does show is the independence of the characteristics expressed in the pseudo-axioms of the second-level theory—a different matter entirely.

Frege's appraisal of Hilbert's enterprise may now be cast in the form of a dilemma: *Either* Hilbert's substitution-maneuver is taken seriously—in which case what Hilbert has constructed is a second-level theory and what he has proved concerns its concepts, not the axioms of Euclidean geometry; *or* Hilbert's claim that he has been concerned with Euclidean geometry and its axioms is taken seriously—in which case the proof-procedure employed is inappropriate and, as Frege's systematic critique shows, Hilbert has not even constructed an axiomatic system. In short, either he has proved something other than what he set out to prove, or what he has argued is nonsense.

Frege concludes, therefore, that Hilbert's attempt did not achieve its avowed aim; and more importantly, that any attempt to prove the independence of first-level axioms in Hilbert's way *cannot* succeed. And having argued thus, he goes on to sketch a way in which Hilbert's aim might be achieved (see pp. 103-12 below). Here he falls back on his notion of thought or sense. Let Ω be a group of true thoughts, and let G be a thought that follows from one or more thoughts of Ω by an inference involving only the laws of logic and the relevant thoughts of Ω. Add G to Ω and calls this procedure *taking a logical step*. If by means of such steps, where each step takes the preceding one as its basis, we reach a group of thoughts containing the thought *A*, then we call *A* dependent on Ω. If this is not possible, we call *A* independent of Ω. The latter will always be the case when *A* is false.

Now imagine a vocabulary where words of the same language stand on opposite sides of a page, and where the sense of an expression or word on the right is different from that of the corresponding word or expression on the left. Imagine further that the vocabulary is constructed in such a way that to a proper name on the right there again corresponds a proper name on the left; to a concept-expression on the right, a concept-expression on the left; and so on. Let this correspondence occur with retention of level, so that only expressions of the same level are opposed. Since each word has its own determinate sense, in producing such a correlation of expressions we have also produced a correlation of senses. Consider the system of propositions expressive of Ω. We can now construct a new system of thoughts Ω′ by the simple expedient of translating the propositions of Ω, using our vocabulary. We do this in such a way that to the premises in Ω, there again correspond premises in Ω′; to the inferences in Ω, there again correspond inferences in Ω′; and so on. Of course, a translation of purely logical expressions will not be required or indeed be possible, for as Frege puts it, logic "brooks no replacement." Suppose, then, that according to our vocabulary, to the thoughts of Ω there corresponds a group of thoughts Ω′ which are true;[34] and that to a thought G of Ω there corresponds a thought G′ of Ω′. If G′ is false, then G is independent of Ω; for if it were dependent, then G′ would have to be dependent on Ω′, and since the latter is a group

34. Strictly speaking, thoughts, not propositions, are true for Frege.

of true thoughts, *G'* would have to be true in virtue of its
dependence.

It must be added that Frege does not consider the preceding
a final and complete description of the proof-procedure, since
several things are lacking:

> In particular, we will find that this basic law which I have
> attempted to elucidate by means of the above-mentioned
> vocabulary still needs more precise formulation, and that
> to give this will not be easy. Furthermore, it will have to
> be determined what counts as a logical inference and what
> is proper to logic. (P. 110 below)

Further:

> If, following the suggestions given above, one wanted to
> apply this to the axioms of geometry, one would still need
> propositions that state, e.g. that the concept *point,* the rela-
> tionship of a point's lying in plane, etc., do not belong
> to logic. These propositions will probably have to be taken
> as axiomatic. Of course, such axioms are of a very special
> kind and cannot otherwise be used in geometry. But here
> we are in unexplored territory (p. 110-11 below).

Part I

Frege, Hilbert, and Korselt

Letter from G. Frege to Heinrich Liebmann

Bad Steben
8.25.1900

Dear Sir:

Since you are interested in the foundations of geometry, I am sending you a copy of my correspondence with Prof. Hilbert about his Festschrift. You may gather from this that it is important to me that you should get to know more precisely my views on this matter. To be sure, the correspondence has already been at a standstill for more than half a year, but I have a card from Prof. Hilbert which promises a continuation. Perhaps a letter is already waiting for me in Jena which my wife, in her well-intentioned way, is holding back. In rereading my letters, I get the impression that in those days I judged Hilbert's investigations more favorably than I do now. But I must await Prof. Hilbert's answer. I suggested to him that because of the importance of the questions with which it dealt, he consider a future publication of our correspondence.

/2/ I now want to try and see whether I can succeed in expounding to you briefly what I understand by second-level concepts. First of all, I must emphasize the radical difference between concepts and objects, which is of such a nature that a concept can never substitute for an object, or an object for a concept. Here it is impossible to give proper definitions. The nature of concepts can be characterized by the fact that they are said to have a predicative nature. An object can never be predicated of anything. When I say, "The Evening Star is Venus," then I predicate not Venus, but *coinciding with Venus*. Linguistically, proper names correspond to objects, concept-words (*nomina appellativa*) to concepts. However, the sharpness of this distinction is somewhat blurred in ordinary language by the fact

From *Gottlob Frege: Kleine Schriften,* ed. I. Angelelli (Hildesheim, Georg Olms Verlagsbuchhandlung, 1967), pp. 404-07.

that what originally were proper names (e.g. "Moon") can become concept-words, and what originally were concept-words (e.g. "god") can become proper names. Concept-words occur with the indefinite article, with words like "all," "some," "many," etc. There occur many other refinements which I shall not investigate here. Now between objects and (first-level) concepts there obtains a relation of subsumption: an object falls under a concept. For example, Jena is a university town. Concepts are generally composed of component-concepts—the characteristics. *Black silken cloth* has the characteristics *black, silken,* and *cloth.* An object falling under this concept has these characteristics as its properties. What /3/ is a characteristic with respect to a concept is a property of an object falling under that concept. Quite distinct from this relation of subsumption is that of the subordination of a first-level concept under a first-level concept, as in "All squares are rectangles." The *characteristics* of the superordinate concept (rectangle) are also *characteristics* of the subordinate one (square). When I say, "There is at least one square root of 4," I am predicating something not of 2 or −2, but of the concept *square root of 4.* Neither am I giving a characteristic of this concept; rather, this concept must already be completely known. I am not singling out any components of this concept, but am stating a certain composition of the concept in virtue of which it differs for example from the concept *even prime number greater than 2.* I compare the individual characteristics of a concept to the stones constituting a house; I compare what is predicated in our proposition to a property of the house, e.g. its spaciousness. Here, too, something is predicated; not, however, a first-level concept, but a concept of the second level. *Square root of 4* relates to there-is-existence in a very similar way in which Jena relates to *university town.* Here we have a relation between concepts; not, however, a relation between first-level concepts, as in the case of subordination, but a relation of a first-level concept to a second-level concept, which is similar to the subsumption of an object under a first-level concept. The first-level concept here plays a role similar to that of an object in the case of subsumption. Here, too, one could speak of subsumption; but this relation, although indeed similar, nevertheless is not the same as that of the subsumption of an object under a /4/ first-level concept. I shall say that a first-level concept falls (not under, but) within a second-level concept. The distinction between concepts of the first and second levels is just as sharp as that between objects and concepts of the first level; for objects can never substitute for concepts. Therefore an object can never fall under a second-level concept; such would be not false but nonsensical. If one tried something like this linguistically, one would get neither a true nor a false thought, but no thought at all. By the way, I once published something about concepts and objects in Avenarius's journal.* A different feature of first-level concepts is expressed by the proposition that if an object falls under such a concept, another object distinct from the preceding one also falls under it. Here we have a second concept of the second level. From both, as second-level characteristics, we can form a third second-level concept within which fall all those first-level concepts under which fall at least two distinct objects. The concepts *prime number, planet,* and *human being* would be such as fall within this second-level concept. It seems to me that Prof. Hilbert initially has the idea of defining second-level concepts; but he does not distinguish them from those of the first level. This explains what always does remain unclear in Hilbert's expositions: how the same concept is apparently defined twice. It isn't the same one at all. At first it is a second-level concept; afterwards it is a first-level concept that falls within the former. One of the mistakes that occur here is to confuse these and to associate the same word (e.g. "point") with both.

With best regards,
sincerely,
G. Frege

---

* Presumably this reference is to "Concept and Object," which appeared in R. Avenarius, ed. *Vierteljahrsschrift für wissenschaftliche Philosophie, 16* (1892). [Trans.]

Frege–Hilbert Correspondence Leading to "On the Foundations of Geometry"

FREGE TO HILBERT

/1/ . . . I begin with a statement by Thomae about your explanation in § 3. He said something like this: "That is not a definition, for no characteristics whatever are mentioned by which an occurrence of the relation of between could be recognized." I, too, cannot think of it as a definition; however, neither do you call it that, but rather an explanation. You apparently use the two expressions "explanation" and "definition" to designate distinct things; but the difference is not clear to us. The explanations in § 4 seem to be of the very same kind as your definitions. For example, we are here told to what the words "lying on the straight line *a* on the same side of the point 0" are supposed to refer; just as in the subsequent definition, for example, we are told to what the word "line-segment" is supposed to refer. The explanations of §§ 1 and 3 are apparently of an entirely different nature: there the references of the words "point," "straight line," "plane," and "between" are not given but rather are presupposed as known. At least it seems that way. However, it also is unclear what you call a point. One first thinks of points in the sense of Euclidean geometry, and is confirmed in this by the proposition that the axioms express

These three letters are from *Gottlob Frege: Kleine Schriften,* pp. 407-18. Dates and places of writing are missing, as are the beginnings and endings of the letters.

basic facts of our intuition. Later on, however (p. 20), you conceive of a pair of numbers as a point. I am uneasy about the propositions (§ 1) that the precise and complete description of relations is provided by the axioms of geometry, and that (§ 3) axioms define the concept "between." Here axioms are saddled with something that is the function of definitions. It seems to me that this blurs the boundaries /2/ between axioms and definitions in a serious manner, and that besides the old reference of the word "axiom" which emerges in this proposition—that axioms express basic facts of intuition—there emerges another which, however, is not quite comprehensible to me. At the present time, confusion regarding definitions is already prevalent in mathematics, and many seem to act according to the maxim:

> If you can't quite give a demonstration,
> Consider it an explanation.

In view of this, I do not approve of increasing this confusion by using the word "axiom" in a vacillating sense, sometimes like "definition." I think it is high time we began to come to an understanding about what a definition is and what it is supposed to accomplish, and what principles are therefore to be followed in giving definitions (my *Basic Laws of Arithmetic, 1,* § 33). It seems to me that at the present time complete anarchy and subjective inclination rule supreme in this area. Permit me to expound some of my thoughts on this subject. I should like to divide the totality of mathematical propositions into definitions and all other propositions (axioms, principles, theorems). Every definition contains a sign (expression, word) which previously has had no reference and which is given a reference only through this definition. Once this has happened, one can make out of this definition a self-evident proposition which /3/ is then to be used like an axiom. But we must adhere to the tenet that in a definition nothing is asserted; rather, something is stipulated. Therefore, what requires a proof or some other reasoning to establish its truth ought never to be presented as a definition. I use the equality-sign as an identity-sign. Let us, then, suppose that the references of the plus-sign, the three-sign, and the one-sign are known; but that the four-sign is unknown. We can then assign a reference to the four-sign by means of the equation "$3 + 1 = 4$." Once this has been done, the equation will be true of itself and no longer needs proof. However, if one wanted to escape the onus of a proof by setting up a definition, this would be logical sleight of hand. It is absolutely essential for the rigor of mathematical investigations that the difference between definitions and all other propositions be maintained throughout in all its sharpness. The other propositions (axioms, principles, theorems) must contain no word (sign) whose sense and reference or (in the case of form-words, letters in formulae) whose contribution to the expression of the thought is not already completely settled, so that there is no doubt about the sense of the proposition—about the thought expressed in it. Therefore it can only be a question of whether this thought is true; and if it is, on what its truth /4/ might rest. Therefore it can never be the purpose of axioms and theorems to establish the reference of a sign or word occurring in them; rather, this reference must already be established. We may assume that there are propositions of yet a third kind: the explicatory propositions, which, however, I should not like to consider as belonging to mathematics itself but instead should like to relegate to the preamble, to a propaedeutic. They resemble definitions in that in their case, too, what is at issue is the stipulation of the reference of a word (sign). Therefore they, too, contain something whose reference at least cannot be presupposed as completely and unquestionably familiar, simply because in the language of ordinary life it is used in a vacillating and equivocal manner. If the reference to be assigned in such a case is logically simple, one cannot give a real definition but must be satisfied with ruling out by means of hints the unwanted references present in ordinary usage, and with indicating those that are intended. Hereby, of course, one always has to count on co-operative understanding, even guessing. Such explicatory propositions cannot be used in proofs, as definitions are. For that, they lack the requisite precision; and because of this I should like to relegate them to the preamble, as I have said. I call axioms propositions that are true but that are not proved because our understanding of them derives from that nonlogical basis which may be /5/ called intuition of space. From the fact that axioms are true, it follows of itself that they do not contradict one another. This needs no further proof. Definitions, too, must not contradict one another. If they do, they are faulty. The principles for giving definitions must be such that when they are observed, no contradiction can occur. If I were to posit your Axiom II.1 as such, I should thereby be presupposing that the references of the expressions "something is a point of a straight line" and "B lies between A and C" are completely and unequivocally known, and that in the latter case I should be presupposing this in general, whatever might be understood by the letters. In which case the axioms cannot serve to give a more precise explanation, for example, of the word "between"; and it is obviously impossible to give this word still another reference later on, as you suggest on p. 20. If this reference is distinct from that of the word "between" in § 3, then you are faced with a most serious equivocation. It seems that nothing is left for us but to assume that in II.1 the word "between" has as yet no reference at all. In which case, however, II.1 cannot be true, and hence cannot be an axiom in my sense of the word, which, I think, is the commonly accepted one. If, as is then probable, that word does not yet have a sense in § 3, then the proposition II.1 also does not have a sense, expresses no thought. Consequently it does not express a basic fact of our intuition either. But then, /6/ what is it supposed to do? Is it perhaps intended to stipulate the reference of "between," like a definition? Then this cannot later be done once again. In § 6 you say, "The axioms of this group define the concept of congruence or of motion." But then, why are they not called "definitions"? Precisely what difference —if any—is there between definitions and axioms? Of course the latter do not fulfill the requirements that must be made of definitions, if for no other reason than that there are several of them; and also because in them there occur expressions ("on a given side of a straight line *a*") whose very references do not yet seem to be established. I do not fail to see that in order to prove the independence of the axioms from one another, you have to place yourself on a higher plane, from which Euclidean geometry appears as no more than a particular case of a more general one. But it seems to me that, for the reasons cited, the course you adopt is not quite feasible. I should not consider your essay to be valuable if I did not believe that I could somehow see how such objections could be met; but this will probably be impossible without major revision. In any case, it seems to me that above all what is necessary is an understanding about the expressions "explanation," "definition," "axiom." Here you seem to deviate considerably from that with which I am familiar and probably also from what is customary, which makes it difficult for me to /7/ keep these expressions apart in your exposition and to understand the logical structure clearly. . . .

Hilbert's Reply to Frege

. . . Another preliminary remark: If we want to understand each other, we must not forget the difference in the nature of the purposes that guide us. I was forced to construct my system of axioms by the following necessity: I wanted to provide an opportunity for understanding those geometric propositions which I consider to be the most important products of geometric investigations—that the axiom of parallels is not a consequence of the remaining axioms; similarly the Archimedean axiom; etc. I wanted to answer the question whether it is possible to prove the proposition that two equal rectangles having the same base line also have equal sides,[1] or whether, as with Euclid, it is a new postulate. In fact, I wanted to create the possibility of understanding and answering such questions as why the sum of the angles of a triangle is two right angles, and how this fact is related to the axiom of parallels. I believe my Festschrift shows that my system of axioms permits us to answer such questions in a very determinate manner, and that it provides quite surprising and indeed quite unexpected answers to many of these questions. Several subsequent investigations by my students—I take the liberty of indicating Mr. Dehn's dissertation, which will appear in the near future in *Math. Annalen*[*]— show the same thing. This, then, was my primary purpose. At the same time I do believe that I have also constructed a system of geometry /8/ that fulfills even the most stringent requirements of logic; and herewith I come to the actual reply to your letter.

You say that my explanation in § 3 is not a definition of the concept "between," for the characteristics are missing. Of course —the characteristics are stated at length in Axioms II.1–II.5. However, if one wants to take the word "definition"[†] precisely, in the customary sense, then one has to say,

> "Between" is a relation of points of a straight line, and has the following characteristics: II.1 . . . II.5.

You go on to say, "Surely the explanation in § 1, where the meanings point, straight line . . . are not stated but presupposed as known, are quite different." I think that here we have the crux of our misunderstanding. I do not want to presuppose anything as known. I think that with my explanation in § 1 I give the definitions of the concepts point, straight line, and plane if one again adds to these all the axioms of axiom-groups I–V

1. After all, this theorem is the basis of all surface measuring.
* *Die Legendreschen Sätze über die Winkelsumme im Dreieck;* appeared in *Mathematische Annalen, 53* (1900), 404-39. [Trans.]
† The German has no quotation marks around 'definition'. [Trans.]

as characteristics. If someone is looking for other definitions of "point," etc., perhaps by means of circumscriptions like extensionless, then of course I would most decidedly have to oppose such an enterprise. One is then looking for something that can never be found because there is nothing there, and everything gets lost, becomes confused and vague, and degenerates into a game of hide-and-seek. If you prefer to call my axioms characteristics of the concept that are posited in the "explanations" and consequently exist, I should have no objections to this except, perhaps, that it contradicts the custom of mathematicians and physicists. Of course I /9/ must also be able to do as I please in the matter of positing characteristics; for as soon as I have posited an axiom, it will exist and be "true." And herewith I come to another important point of your letter. You write, "Axioms I call propositions . . . From the fact that axioms are true it follows that they do not contradict one another." I was extremely interested to read just this proposition in your letter, because for as long as I have been thinking, writing, and lecturing about such things, I have always been saying the opposite: If the arbitrarily posited axioms together with all their consequences do not contradict one another, then they are true and the things defined by these axioms exist. For me, this is the criterion of truth and existence. Either the proposition "Every equ.[ation] has a root" is true, or the existence of the roots is proved as soon as the axiom "Every equ. has a root" can be added to the remaining arithmetic axioms without it being the case that there can ever arise a contradiction, whatever inferences may be drawn. This conception is of course the key not only to the understanding of my Festschrift, but also e.g. to my talk in Munich about the axioms of arithmetic, where I proved or at least indicated that the system of all common real numbers exists, whereas the system of all of Cantor's powers or even of all alephs does not exist—as Cantor himself asserted in a similar sense, using scarcely different words. Therefore, to reiterate the main point: The change of name—

"characteristic" instead of "axiom," etc.—is surely a mere super-ficiality and moreover a matter of taste; and in any case, it is easily accomplished. On the other hand, in my estimation it /10/ is impossible to give a definition of a point in 3 lines, since it is only the whole axiom-structure that gives the complete definition. After all, each axiom contributes something to the definition, and therefore each new axiom alters the concept. "Point" is always something different in Euclidean, non-Euclidean, Archimedean, and non-Archimedean geometry respectively. Once a concept has been completely and unequivocally fixed, then in my opinion the addition of any axiom whatever is entirely impermissible and illogical—a mistake that is made very frequently, especially by physicists. In theoretical physical investigations, sheer nonsense frequently results because in the course of the investigation they continue to construct new axioms that are never confronted with the prior assumptions and of which it is never shown whether they contradict any of the facts that follow from the previously constructed axioms. It is precisely this procedure of constructing an axiom, appealing to its truth (?), and concluding from this that it is compatible with the defined concepts, that is a primary source of mistakes and misunderstandings in modern physical investigations. A primary purpose of my Festschrift was to avoid these mistakes.

There is only one further objection which I must touch upon. /11/ You say that my concepts, e.g. "point," "between," are not unequivocally fixed; that on p. 20, for example, "between" is taken in different senses and that there a point is a pair of numbers. — But surely it is self-evident that every theory is merely a framework or schema of concepts together with their necessary relations to one another, and that the basic elements can be construed as one pleases. If I think of my points as some system or other of things, e.g. the system of love, of law, or of chimney sweeps . . . and then conceive of all my axioms as relations between these things, then my theorems, e.g. the Pythagorean one, will hold of these things as well. In other words, each and every theory can always be applied to infinitely many systems of basic elements. For one merely has to apply a univocal and reversible one-to-one transformation and stipulate that the axioms for the transformed things be correspondingly similar. Indeed, this is frequently applied, for example in the principle of duality, etc.; I also apply it in my independence-proofs. The totality of assertions of a theory of electricity does of course hold of every other system of things substituted in place of the concepts magnetism, electricity, . . . just as long as the required axioms are fulfilled. However, the state of affairs just indicated can never be a shortcoming[2] of a /12/ theory, and in any case is unavoidable. Of course, in my opinion at least, the application of a theory to the world of appearances always requires a certain measure of goodwill and tact: that for points, one substitute bodies as small as possible; for straight lines, lines as long as possible, perhaps lightrays; etc. Nor can one be too exacting in examining the propositions, for after all, they are only propositions of the theory. However, the more developed a theory is and the more ramified its structure, the more self-evident will be the manner of its application to the world of appearances; and it would require a large measure of bad intentions indeed if one wanted to apply the more precise propositions of plane geometry or Maxwell's theory of electricity to appearances other than the ones for which they were intended. . . .

FREGE'S REPLY TO HILBERT

. . . I believe that from the paper you gave in Munich—thank you very much for sending me a copy—I have understood your plan still a little better. It seems to me that you want to divorce geometry completely from our intuition of space and make it a purely logical discipline, like arithmetic. If I understand you correctly, the axioms that are no doubt usually considered the basis of the whole structure on the assumption that they are

2. On the contrary: it is a tremendous advantage.

guaranteed by the intuition of space, are to be carried as conditions in every theorem; /13/ not, to be sure, in their complete text, but as included in the words "point," "straight line," etc. They are intended to prove the mutual independence and consistency of certain propositions (axioms): the unprovability of propositions from certain presuppositions (axioms). Considered purely logically, all this amounts to one and the same thing: It is intended to show the consistency of certain determinations. "D is not a consequence of A, B, C" says the same thing as "The occurrence of A, B, C does not stand in contradiction with the nonoccurrence of D." "A, B, C are independent of one another" means "C is not a consequence of A and B, B is not a consequence of A and $C,$ and A is not a consequence of B and C." After everything has thus been reduced to the same schema, this question arises: What means do we have for proving that certain properties or requirements (or however else one wants to put it) do not contradict one another? The only way I know of is to present an object that has all of these properties, to exhibit a case where all these requirements are fulfilled. Surely it is impossible to prove consistency in any other way. Therefore, if it is a matter of proving the mutual independence of the axioms, it will have to be shown that the nonoccurrence of one of these axioms does not stand in contradiction with the occurrence of the remainder. (I here accommodate myself to your usage of "axiom.") Now it will be impossible to adduce such an example /14/ in the realm of elementary Euclidean geometry, simply because all its axioms are true. In that you place yourself on a higher plane from which Euclidean geometry appears to be but a particular case of a more inclusive theory, there arises the expectation of examples that make this mutual independence of the axioms obvious. To be sure, at this point some misgivings arise for me, but I do not want to pursue it any further here. A main point seems to me to be that you want to consider Euclidean geometry from a higher level. And indeed, either it is possible to prove the mutual independence of the axioms in this way, or it will not be provable at all. And I do think that such an undertaking, if it is directed toward the axioms in the traditional sense of Euclidean geometry, is also of the greatest scientific interest. The scientific significance of such an investigation would generally be much less if it covered a system of arbitrarily posited propositions. Because of the misgivings indicated above, I don't dare decide whether it is possible to prove the mutual independence of the axioms of Euclidean geometry in this way. At any rate, your idea of considering Euclidean geometry as a special case of a more inclusive theory is valuable even without this accomplishment.

I quite agree with you that the genetic method lacks complete logical certainty. It lies in the nature of the case that the development of science has followed this course. In /15/ the meantime, it must not be forgotten that what must be kept in mind as the goal toward which the development aims is to be a logically perfect system.

You write, "Once a concept has been completely and unequivocally fixed, then in my opinion the addition of any axiom whatever is entirely impermissible and illogical." If I understand your opinion correctly here, I can only agree with pleasure; all the more so since I thought myself alone in holding this view. Indeed, the shortcoming of the genetic method probably lies in precisely this, that the concepts are not finished but nevertheless are used in this unfinished and therefore really unusable condition, so that one never knows whether a concept is definitely finished. Thus it happens that one can prove propositions which then are falsified once more by a subsequent development, because different thoughts come to be contained in them. It is precisely such alterations that are dangerous because one never really becomes quite conscious of them, since the same wording is retained.

I also agree with you in your low estimate of the definition of "point" by means of circumlocutions using "extensionless."

However, I should not shrink from the admission that "point" cannot be defined at all.

Therefore it seems that as long as we remain on the level of generalities, we are in satisfactory agreement. The situation changes when we proceed to the actual execution. As I recall, /16/ I had previously considered the possibility of understanding your axioms as components of your explanations; and yet I was surprised to learn that all of the axioms of Groups I–V are to be added to the explanation in § 1 in order to complete it. According to this, then, the explanation and what belongs to it actually take up all of your chapter 1, and many other explanations and theorems are encapsulated in it. I confess that I find this logical edifice puzzling and opaque in the highest degree. During the course of my thinking about defining, I have become increasingly more strict in what I demand from a definition and apparently have removed myself so far from the opinion of most mathematicians that mutual understanding has been made very difficult. What I object to will probably become clearest through a consideration of your explanation in § 3. For comparison, I adduce Gauss's definition of number-congruence. If one knows what difference is and what it means to say that "a number divides evenly into a number," then, owing to this explanation, one is wholly capable of handling the concept of number-congruence and can immediately decide whether, for example, 2   8 (mod. 3). The situation would be completely different if the word "congruent" were to be explained not only by what is known, but also by itself; if, following your lead, one were to say,

> EXPLANATION: Whole numbers stand in certain /17/ relations to one another, and this relation we describe by using above all the word "congruent."

> AXIOM 1. Every number is congruent to itself under any modulus whatever.

AXIOM 2. If a number is congruent to a second, and the latter is congruent to a third under the same modulus, then the first is also congruent to the third under the same modulus. Etc.

Could one gather from such a definition that $2 \equiv 8 \pmod{3}$? Hardly! Matters stand even more unfavorably as regards your explanation of between, for your axioms of ordering also contain the words "point" and "straight line," whose references are also unknown. Your system of definitions resembles a system of equations with several unknowns, where the solvability and particularly the univocity of the determination of the unknowns remains doubtful. If the latter did obtain, then it would be better to give these solutions, i.e. to explain each of the expressions "point," "straight line," "between," separately by means of what is already known. I do not know how, given your definitions, I could decide the question of whether my pocket watch is a point. Already the first axiom deals with two points; therefore, if I wanted to know whether it held of my pocket watch, I should first of all have to know of another object that it is a point. But even if I knew this, e.g. of /18/ my fountain pen, I should still be unable to decide whether my watch and my fountain pen together determine a straight line, because I should not know what a straight line is. Furthermore, the word "determine" would also occasion difficulties. But even if I were to understand the words "point" and "straight line" as in elementary geometry and were given three points on a straight line, given your explanations and the axioms belonging with them I should nevertheless be unable to decide which of these points lies between the other two. I should not even know what investigations I might have to conduct to this purpose. Add to this the following: According to I.7, there are at least two points on every straight line. What would you say to the following?

EXPLANATION: We conceive of objects which we call gods.

AXIOM 1. Every god is omnipotent.

AXIOM 2. Every god is omnipresent.

AXIOM 3. There is at least one god.

Here my distinction between concepts of the first and second levels becomes relevant. I say "my" because I am not aware that it has been drawn with sufficient sharpness before me. In the "there is," we have a concept of the second level which, together with omnipotent and omnipresent, which are of the first level, may be taken as a characteristic of a first-level concept (see my *Foundations of Arithmetic,* § 53, where instead of "level," I said "order"; and my *Basic Laws of Arithmetic,* §§ 21 and 22). The characteristics which you state in your axioms undoubtedly are all of a higher level than the first. That is, /19/ they do not provide an answer to the question, "What property must an object have to be a point, a straight line, a plane, etc.?" Instead they contain, for example, second-level relations, such as that of the concept "point" to the concept "straight line." It seems to me that you really intend to define second-level concepts, but that you don't distinguish them clearly from those of the first level. The questionable nature of the definition of magnitude, which for example Stolz gives in the introduction to his *Lectures on General Arithmetic,* surely stems from the same source. Hereby it is always characteristic that the word "similar" occurs with a totally blurred reference. This can be avoided only by putting the question in a completely different way. And the repair of the defects which I believe I see in your definitions will probably also have to be made in a similar manner. Only in this case it will be much more difficult, because instead of one system we have three (points, straight lines, and planes), and many different relations enter in. Moreover, what are you calling a system here? I believe it is the same as what is usually called a set or class, and what is most correctly called the extension of a concept.

Our views are probably most sharply opposed with respect to your criterion of existence and truth. But perhaps I don't understand your opinion correctly. To become clear about this, I submit the following example: Let us suppose that we know that the propositions

/20/ 1. *A* is an intelligent being
2. *A* is omnipresent
3. *A* is omnipotent

together with all their consequences did not contradict one another. Could we infer from this that there exists an omnipotent, omnipresent, intelligent being? I don't see how! The principle would go something like this:

If (generally, whatever *A* may be) the propositions

A has the property ϕ
A has the property Ψ
A has the property χ

together with all their consequences do not contradict one another, then there exists an object that has all of the properties ϕ, Ψ, χ.

This principle is not at all evident to me; and if it were true, it would probably be useless. Are there any means of proving consistency other than that of exhibiting an object that has all of the properties? However, if one has such an object, one would not need to prove that there is one by the roundabout way of proving consistency.

If a general proposition contains a contradiction, then every particular proposition included under it will do likewise. Therefore from the consistency of the latter we can infer that of the general one [proposition], but not vice versa. Suppose we had /21/ proved that in a right triangle with two equal sides, the square on the hypotenuse is twice as large as that on one of the other two sides—which is easier than proving the general Pythagorean theorem. From this we can now infer that the proposition contains no contradiction; that the square on the hypotenuse of a right triangle having two equal sides is equal to the sum of the squares on the other two sides. From this we can deduce further that the general Pythagorean theorem contains no contradiction. But can we conclude still further that the Pythagorean theorem is therefore true? I cannot admit such an inference from consistency to truth. Presumably you don't mean it that way either. In any case, a more precise formulation seems necessary.

Furthermore, it seems to me that a logical danger lies in the fact that you say, for example, "the axiom of parallels," as if it were the same in every other geometry. Only the wording is the same; the thought-content is different in each particular geometry. It would not be correct to call the previously mentioned particular case of the Pythagorean theorem *the* Pythagorean theorem; for after one has proved that particular case, one still has not proved the Pythagorean theorem. Even if it is supposed that these axioms in the particular geometries all are particular cases of general axioms, then although from the consistency in a particular geometry one could indeed infer the consistency in the general case, nevertheless one could not infer consistency in other particular cases.

Concerning what you say about the applicability of a /22/ theory and about the univocal and reversible one-to-one transformation, I reserve the right to make a reply . . .

On the Foundations of Geometry

BY GOTTLOB FREGE IN JENA

I.

Mr. Hilbert's Festschrift concerning the foundations of geometry[1] prompted me to write to the author, setting forth my own divergent views; and out of this grew an exchange of letters which unfortunately was soon terminated. Believing that the questions dealt with herein might be of more general interest, I contemplated its future publication. However, Mr. Hilbert has some reservations about agreeing to this, since in the meantime his own views have changed. I regret this stand, since by means of this correspondence the reader would most conveniently have been familiarized with the state of the question, and I would have been spared a new composition. However, it seems to me that the views in this area are still so divergent and still so far removed from any clarification that a public discussion for the purpose of bringing about an understanding would be quite justified. Therefore I should here like to consider some questions of fundamental importance, and I should like to do so in the form of a discussion of Mr. Hilbert's essay. And for this purpose it may be irrelevant whether at present the honored author still maintains those of his views that are being questioned here.

From *Jahresbericht der deutschen Mathematiker-Vereinigung,* *12* (1903), 319-24, 368-75.
1. Festschrift for the festival marking the unveiling of the Gauss-Weber Memorial in Göttingen (Leipzig, Teubner, 1899).

To begin with, let us deal with these questions: What is an axiom? What is a definition? In what relations might these stand to one another?

Traditionally, what is called an axiom is a thought whose truth is certain without, however, being provable by a chain of logical inferences. Logical laws, too, are of this nature. Some people may nevertheless be inclined to refrain from ascribing the name "axiom" to these general laws of inference, but rather wish to reserve it for the basic laws of a more restricted field, e.g. geometry. But this is a question of less consequence. Here we shall not go into the question of what might justify our taking these axioms to be true. In the case of geometrical ones, intuition is generally given as a source.

In mathematics, what is called a definition is usually the stipulation of the reference of a word or sign. A definition /320/ differs from all other mathematical propositions in that it contains a word or sign which hitherto has had no reference, but which now acquires one through it. All other mathematical propositions (axiomatic ones and theorems) must contain no proper name, no concept-word, no relation-word, no function-sign, whose reference has not previously been established.[2] Once a word has been given a reference by means of a definition, we may form self-evident propositions from this definition, which may then be used in constructing proofs in the same way in which we use principles.[3] For example, let us suppose that the references of the plus-sign, the three-sign, and the one-sign are known; we can then assign a reference to the four-sign by means

2. With few exceptions ('π', 'e'), letters do not, as a rule, have a reference; they do not designate anything, but only indicate in order to lend generality to the thought. As with certain form-words, we cannot require a reference from them; but the manner in which they contribute to the expression of the thoughts must be definite. I have given a protracted treatment of the usage of letters in my *Basic Laws of Arithmetic, 1* (Jena, Pohle, 1893), §§ 8, 9, 17, 24, 25.

3. What I here call a principle is a proposition whose sense is an axiom.

of the definitional equation "3 + 1 = 4." Once this has been done, the content of this equation is true of itself and no longer needs proof. Nevertheless, it would be inappropriate to count definitions among principles. For to begin with, they are arbitrary stipulations and thus differ from all assertoric propositions. And even if what a definition has stipulated is subsequently expressed as an assertion, still its epistemic value is no greater than that of an example of the law of identity $a = a$. By defining, no knowledge is engendered; and thus one can only say that definitions that have been altered into assertoric propositions formally play the role of principles but really are not principles at all. For although one could just possibly call the law of identity itself an axiom, still one would hardly wish to accord the status of an axiom to every single instance, to every example, of the law. For this, after all, greater epistemic value is required. No definition extends our knowledge. It is only a means for collecting a manifold content into a brief word or sign, thereby making it easier for us to handle. This and this alone is the use of definitions in mathematics.[4] Never may a definition strive for more. And /321/ if it does, if it wants to engender real knowledge, to save us a proof, then it degenerates into logical sleight of hand. In the case of some of the definitions which one finds in mathematical writings, one should like to write in the margin,

> If you can't quite give a demonstration,
> Consider it an explanation.

Never may something be represented as a definition if it requires proof or intuition to establish its truth. On the other hand, one

4. One might also represent as a use of a definition that through it one becomes more clearly aware of the content of what one has connected, albeit only half-consciously, with a certain word. This may occur but is less a use of the definition than of defining. Once a definition has been set up, it is irrelevant for what follows whether the explained word or sign has just been newly invented, or whether previously some sense or other had been connected with it.

can never expect principles or theorems to settle the reference of a word or sign. It is absolutely essential for the rigor of mathematical investigations, not to blur the distinction between definitions and all other propositions.

Axioms do not contradict one another, since they are true; this does not stand in need of proof. Definitions must not contradict one another. We must set up such guidelines for giving definitions, that no contradiction can occur. Here it will essentially be a matter of preventing multiple explanations of one and the same sign.[5] The usage of the words "axiom" and "definition" as presented in this paper is, I think, the traditional and also the most expedient one.

As to Mr. Hilbert's Festschrift, it confronts us with a peculiar confusion of usage. When it is said in the introduction, "Geometry requires . . . for its consequential construction only a few simple basic facts. These basic facts are called axioms of geometry," then this is quite in keeping with what has just been set forth; similarly when it is said in § 1, p. 4, "The axioms of geometry fall into five groups; each one of these groups expresses certain basic and interconnected facts of our intuition."[6]

A completely different view appears to lie at the basis of this pronouncement (§ 3): "The axioms of this group define the concept 'between'." How can axioms define something? Here axioms are saddled with something that is the function of definitions. The same remark obtrudes itself when in § 6 we read, "The axioms of this group define the concept of congruence or of motion."

/322/ Due to Mr. Hilbert's kindness I am now in a position to say in what sense he has used the word "axiom." For him, the axioms are components of his definitions.[7] So, for example,

5. Compare my *Basic Laws of Arithmetic, 2,* §§ 56-57.
6. In the first of the propositions quoted, the axioms are thoughts, to be sure; in the second, they are expressions of thoughts: propositions.
7. The time at which he held this view must be assumed to be that of the writing of the Festschrift and of the date of his letter (12.29.99).

axioms II.1 to II.5 are components of the definition of *between*.
Between, therefore, is a relation of those points of a straight
line to which axioms II.1 to II.5 apply. In the Festschrift, II.1
reads like this:

> If *A, B, C* are points of a straight line and *B* lies between
> *A* and *C,* then *B* also lies between *C* and *A.*

The axioms state the characteristics that would otherwise be
missing from the explanations. Similarly, the explanation in § 1
of the Festschrift also contains the definitions of the concepts
point, straight line, and plane if one adds to it all of the axiom-
groups I to V, whose presentation takes up the whole first chapter. The first definition, then, extends thus far. Other definitions
are encapsulated in it, for example that of *between;* as well as
theorems, for example congruence-theorems. On the basis of
this it is not altogether easy to see which parts of the first chapter belong to that definition. At least it is difficult to believe that
the theorems should also be considered as such components.
This explains Mr. Hilbert's statement that axioms define something. But is it compatible with this, that axioms express basic
facts of our intuition? If they do, then they assert something.
But then, every expression that occurs in them must already
be fully understood. However, if axioms are components of
definitions, then they will contain expressions such as "point"
and "straight line" whose references are not yet settled but are
still to be established. And then each single axiom is something
dependent, something that cannot be thought without the other
axioms that belong to the very same definition. It is only on p. 19
of the Festschrift that the reference of the word "point" is established according to Mr. Hilbert's intentions. It is only now that
the axioms presented so far express thoughts that are true in
virtue of the definition; but for that very reason they do not
express basic facts of our intuition, since then their validity
would be based precisely on this intuition. Let us take the fol-

lowing simple example. We may rewrite the definition "A rectangle is a parallelogram with a right angle" thus:

> EXPLANATION: Conceive of plane figures which we call
> rectangles.

> AXIOM 1. All rectangles are parallelograms.

> /323/ AXIOM 2. In every rectangle there is a pair of sides that
> stand perpendicular to each other.

These two axioms must be regarded as inseparable components of the explanation. If, for example, we were to leave
out the first axiom, then the word "rectangle" would acquire
a different reference; and if upon completing the definition we
went on to posit the remaining second axiom as an assertoric
proposition, it too would thereby acquire a different sense from
the one it now has through its connection with the first. That is,
it would not even be the same proposition; not, at least, if one
considers the thought expressed in it essential to the proposition.

Once the explanation including the two axioms has been
posited, the latter may be asserted as true; however, their truth
will not be founded on an intuition, but on the definition. And it
is precisely because of this that no real knowledge is contained
in them—something which undoubtedly is the case with axioms
in the traditional sense of the word.

Now in chapter 2, Mr. Hilbert considers the questions whether
or not the axioms contradict one another, and whether or not
they are independent of one another. Now, how is this independence to be understood? After all, each of the two axioms
needs the other just to be what it is. Similarly in other cases.
It is only through all of the axioms that according to Mr. Hilbert belong to, for example, the definition of a point, that the
word "point" acquires its sense; and consequently it is also
only through the totality of these axioms that each single axiom
in which the word "point" occurs acquires its full sense. A sep-

aration of the axioms in such a way that one considers some as valid and others as invalid is inconceivable because thereby even those that are taken to be valid would acquire a different sense. Those axioms that belong to the same definition are therefore dependent on each other and do not contradict one another; for if they did, the definition would have been postulated unjustifiedly. However, neither can one investigate before they are postulated, whether these axioms contradict one another, since they acquire a sense only through the definition. There simply cannot be any question of contradiction in the case of senseless propositions.

How, then, are we to understand Hilbert's formulation of the question? We may assume that it does not concern the whole axioms[8] but only those of their parts that express characteristics of the concepts to be defined. In the case of our example, the characteristics /324/ are *parallelogram* and *having two sides standing perpendicular to each other*. If these did contradict one another, no object having these two properties could be found; in other words, there would be no rectangle. Conversely: if one can produce a rectangle, then this means that these characteristics do not contradict one another; and in fact this is just about the way in which Mr. Hilbert proves the consistency of his axioms. In reality, however, this is merely a matter of the consistency of the characteristics. Similarly concerning independence. If from the fact that an object has a first property it may generally be inferred that it also has a second, then one may call the second dependent upon the first. And if these properties are characteristics of a concept, then the second characteristic is dependent upon the first. This is just about the way in which Mr. Hilbert proves the independence of his axioms (more correctly, of the characteristics). For the time being, the matter may be thought of in this way. And yet, it really is not quite

8. As one can see, here, as in the preceding, I accommodate myself to Hilbert's usage.

as simple as it may appear to be according to the preceding. If we want to get to the bottom of this, we shall have to consider the peculiarities of Hilbert's definitions more closely; and that will be done in a subsequent essay.

II.

/368/ Mr. Hilbert's definitions and explanations appear to be of two kinds. The first explanation of § 4 explains the expressions that points lie in a straight line and on the same side of a point, and that points lie in a straight line but on different sides of a point. Once the expressions "point of a straight line *a*" and "a point lies between a point *A* and a /369/ point *B*" are understood, then given this explanation, one knows precisely to what the explained expressions refer. The explanation of § 9 is of an entirely different kind. Here we read:

> The points of a straight line stand in a certain relation to one another which we describe by using above all the word *"between."*

From this we obviously do not get to know the reference of the word "between." However, the explanation is still incomplete. It is to be completed by the following axioms:

> II.1. If *A, B, C* are points of a straight line, and *B* lies between *A* and *C,* then *B* also lies between *C* and *A*.

> II.2. If *A* and *C* are two points of a straight line, then there is always at least one point *B* that lies between *A* and *C,* and at least one point *D* such that *C* lies between *A* and *D*.

> II.3. For any three points of a straight line there is always one and only one point that lies between the other two.

> II.4. Any four points *A, B, C, D* of a straight line can always be ordered in such a way that *B* lies between *A* and

C as well as between A and D, and also that C lies between A and D as well as between B and D.

But do we learn from this, when the relation of lying-in-between obtains? No, rather the reverse: we recognize the truth of the axioms once we have grasped this relation. If we posit the Gaussian definition of the congruence of numbers, we can easily decide whether 2 and 8 are congruent modulo 3, or what investigations we must conduct in order to find out. All we have to know are the expressions that occur in the definition ("difference," "a number divides evenly into a number"). Now with this let us compare the following explanation, which has been constructed according to Hilbert's pattern:

> Whole numbers stand in certain relations to each other which we describe by using above all the word "congruent."
>
> AXIOM 1. Every number is congruent to itself under any modulus whatever.
>
> AXIOM 2. If a number is congruent to a second, and the latter is congruent to a third under the same modulus, then the first is also congruent to the third under this modulus.
>
> AXIOM 3. If a first number is congruent to a second, and a third is congruent to a fourth under the same modulus, then the sum of the first and third is also congruent to the sum of the second and fourth under this modulus.

And so on.

/370/ Could one gather from such a definition that 2 is congruent to 8 modulo 3? Hardly! And here matters lie still more favorably than in the case of Mr. Hilbert's definition, in which occur the words "point" and "line," whose references are as yet unknown to us. But even if we were to understand these words in the sense of Euclidean geometry, given our definition we could not decide which of three points lying in a straight line lies between the other two.

If we survey the total of Mr. Hilbert's explanations and axioms, it seems comparable to a system of equations with several unknowns; for as a rule, an axiom contains several unknown expressions such as "point," "straight line," "plane," "lie," "between," etc.; so that only the totality of axioms, not single axioms or even groups of axioms, suffices for the determination of the unknowns. But does even the total suffice? Who says that this system is solvable for the unknowns, and that these are uniquely determined? If a solution were possible, what would it look like? Each of the expressions "point," "straight line," etc. would have to be explained separately in a proposition in which all other words are known. If such a solution of Hilbert's system of definitions and axioms were possible, it ought to be given; but surely it is impossible. If we want to answer the question whether an object, for example my pocket watch, is a point, then in the case of the first axiom[9] we are already faced with the difficulty that here two points are being talked about. Therefore we should already have to know an object as a point in order to decide the question of whether my pocket watch together with this point determines a straight line. Not only that: we should also have to know how to understand the word "determine" and what a straight line is. This axiom, then, gets us no further. And so it goes with every one of these axioms; and when we have finally arrived at the last one, we still do not know whether these axioms apply to my pocket watch in such a way that we are justified in calling it a point. Equally as little do we know what sorts of investigations would have to be conducted to decide this question.

In axiom I.7 it is said, "There are at least two points on every straight line." With this, compare the following:

9. "Two distinct points, A and B, always determine a straight line a."

EXPLANATION: We conceive of objects which we call gods.

AXIOM 1. Every god is omnipotent.

AXIOM 2. There is at least one god.

/371/ If this were admissible, then the ontological proof for the existence of God would be brilliantly vindicated. And herewith we come to the crux of the matter. Whoever has seen quite clearly the error contained in this proof will also be aware of the fundamental mistake in Hilbert's definitions. It is that of confounding what I call first- and second-level concepts. I was probably the first to draw this distinction in all its sharpness; and at the time he was writing his Festschrift, Mr. Hilbert evidently was not yet familiar with my papers on this topic.[10] And undoubtedly many others will still be in that position. On the other hand, since without this distinction a deeper insight into mathematics and logic is impossible, I shall try to indicate briefly what this is all about.

Take the proposition "Two is a prime number." Linguistically we distinguish here between a subject, "two," and a predicative constituent, "is a prime number." One usually associates an assertive force with the latter. However, this is not necessary. When an actor on the stage utters assertoric propositions, surely he does not really assert anything, nor is he responsible for the truth of what he utters. Let us therefore remove its assertive force from the predicative part, since it does not necessarily belong to it! Even so, the two parts of the proposition are still essentially different; and it is important to realize that this dif-

10. *The Foundations of Arithmetic. A logico-mathematical investigation into the concept of number.* (Breslau, Köbner, 1884), § 53, where instead of "level," I said "order." "Function and Concept," a paper presented at the 6 January 1891 session of the Society for Medicine and Natural Sciences of Jena (Pohle, 1891), p. 26. *The Basic Laws of Arithmetic, derived in a conceptually perspicuous notation.* (Jena, Pohle, 1893), *1,* §§ 21 ff.

ference cuts very deep and must not be blurred. The first constituent, "two," is a proper name of a certain number; it designates an object, a whole that no longer requires completion.[11] The predicative constituent "is a prime number", on the other hand, does require completion and does not designate an object. I also call the first constituent saturated; the second, unsaturated. To this difference in the signs there of course corresponds an analogous one in the realm of references: to the proper name there corresponds the object; to the predicative part, something I call a concept. This is not supposed to be a definition; for the decomposition into a saturated and an unsaturated part must be considered a logically primitive phenomenon which must simply be accepted and cannot be reduced to something simpler. /372/ I am well aware that expressions like "saturated" and "unsaturated" are metaphorical and only serve to indicate what is meant—whereby one must always count on the cooperative understanding of the reader. Nevertheless, it may perhaps be made a little clearer why these parts must be different. An object, e.g. the number 2, cannot logically adhere to another object, e.g. Julius Caesar, without some means of connection. This, in turn, cannot be an object but rather must be unsaturated. A logical connection into a whole can come about only through this, that an unsaturated part is saturated or completed by one or more parts. Something like this is the case when we complete "the capital of" by "Germany" or "Sweden"; or when we complete "one-half of" by "6."[12]

Now it follows from the fundamental difference of objects

11. Propositions including "all," "every," "some" are of a completely different nature and will not be considered here.
12. From the linguistic point of view, what is to be considered the subject is determined by the form of the proposition. The situation is different when considered from the logical point of view. We may decompose the proposition '8 = 2³' either into '8' and "is the third power of 2," or into '2' and "is something whose third power is 8," or into '3' and "is something which, when the third power of 2, yields 8."

from concepts that an object can never occur predictatively or unsaturatedly; and that logically, a concept can never substitute for an object.[13] One could express it metaphorically like this: There are different logical places; in some only objects can stand and not concepts, in others only concepts and not objects.

Let us now consider the proposition "There is a square root of 4"! Clearly we are here not talking about a particular square root of 4 but rather are concerned with the concept. Here, too, the latter has preserved its predicative nature. For of course instead of the preceding, one can say /373/, "There is something which is a square root of 4," or "It is false that whatever *a* may be, *a* is not a square root of 4." In this case, of course, we cannot divide the proposition in such a way that one part is this unsaturated concept and the other is an object. If we compare the proposition "There is something that is a prime number" with the proposition "There is something that is a square root of 4," we recognize a common constituent: "there is something that." It contains the assertion proper, whereas the

13. In § 49 of his book, *The Principles of Mathematics, 1* (Cambridge, 1903), Mr. B. Russell does not want to concede that a concept is essentially different from an object; concepts, too, are always supposed to be *terms*. He supports his argument here with the contention that we find it necessary to use a *concept* substantively as a *term* if we want to say anything about it, e.g. that it is not a *term*. In my opinion, this necessity is grounded solely in the nature of our language and therefore is not a properly logical one. But at the bottom of p. 508, Mr. Russell once more appears to incline to my opinion. I have treated of this difficulty in my essay, "On Concept and Object," *Vierteljahrsschrift für wissenschaftliche Philosophie, 16* (1892), 192-205. It is clear that we cannot present a concept as independent, like an object; rather it can occur only in connection. One may say that it can be distinguished within, but that it cannot be separated from the context in which it occurs. All apparent contradictions that one may encounter here derive from the fact that we are tempted to treat a concept like an object, contrary to its unsaturated nature. This is sometimes forced upon us by the nature of our language. Nevertheless, it is merely a linguistic necessity.

constitutents that they do not have in common, their predicative and unsaturated nature notwithstanding, play a role analogous to that of the subject in other cases. Here something is asserted of a concept. But clearly there is a great difference between the logical place of the number 2 when we assert of the latter that it is a prime number, and the concept prime number when we say that there is something that is a prime number. Only objects can stand in the former place; only concepts in the latter. Not only is it linguistically inappropriate to say "there is Africa" or "there is Charlemagne"; it is also nonsensical. We may indeed say, "there is something which is called Africa," and the words "is called Africa" signify a concept. The *there is something which,* therefore, is also unsaturated, but in a manner quite different from that of *is a prime number.* In the former case, completion can occur only through a concept; in the latter, only through an object. We take the similarity and the difference of the two cases into account by means of the following mode of expression: In the proposition "2 is a prime number" we say that an object (2) falls *under* a first-level concept (prime number); whereas in the proposition "there is a prime number" we say that a first-level concept (prime number) falls *within* a certain second-level concept. First-level concepts can therefore stand in a relation to second-level concepts that is similar to the one in which objects can stand to first-level concepts.

What applies to concepts also applies to characteristics; for the characteristics of a concept are concepts that are logical parts of the latter. Instead of saying "Two is a square root of 4, and 2 is positive," we may say, "Two is a positive square root of 4"; and we have two component concepts—*is a square root of 4* and *is positive*—as characteristics of the concept *is a positive square root of 4.* We may also call these properties of the number 2 and accordingly say: A characteristic of a concept is a property an object must have if it is to fall under that concept. We have something analogous in the case of second-level concepts. It is easy to infer from the above that first-level concepts can have only first-level characteristics and that second-level concepts can have only second-level characteristics. /374/ A mixture of characteristics of the first and second levels is impossible. This follows from the fact that the logical places for concepts are unsuitable for objects, and that the logical places for objects are unsuitable for concepts. From this it follows further that the explanation beginning with the words "conceive of objects which we call gods" is inadmissible; for the characteristic contained in the first axiom is of the first level, whereas a characteristic of the second level is given in the second axiom.

Now, how do things stand with Mr. Hilbert's definitions? Apparently every single point is an object. From this it follows that the concept of a point (*is a point*) is of the first level, and consequently that all of its characteristics must be of the first level. If we now go through Hilbert's axioms, considering them as parts of the definition of a point, we find that the characteristics stated in them are not of the first level. That is, they are not properties an object must have in order to be a point. Rather, they are of the second level. Therefore, if any concept is defined by means of them, it can only be a second-level concept. It must of course be doubted whether any concept is defined at all, since not only the word "point" but also the words "straight line" and "plane" occur. But let us disregard this difficulty and assume that through his axioms, Mr. Hilbert has defined a concept of the second level. No doubt the relationship of the Euclidean point-concept, which is of the first level, to Hilbert's concept, which is of the second level, will then have to be expressed by saying that according to the convention we adopted above, the former falls within the latter. It is then conceivable—in fact probable—that this does not apply to the Euclidean point-concept alone. And this agrees with what is said on p. 20 [of the Festschrift]: "Consider a pair of numbers (x,y) of the domain Ω to be a point," etc. If previously the word "point" had already been given a reference by means of the definition and the axiom belonging to it, then at this juncture it could not be defined once again. We should probably construe the matter thus: The first-level concept *is a pair of numbers of the domain* Ω, just as the Euclidean concept of a point, is supposed to fall within Hilbert's second-level concept (in case there is one). The use of the word "point" in both cases is, of course, irritating; for obviously the word has distinct references in the two cases.

According to the preceding, Euclidean geometry presents itself as a special case of a more inclusive system which allows for innumerable other special cases—innumerable geometries, if that word is still admissible. And in every one of these /375/ geometries there will be a (first-level) concept of a point and all of these concepts will fall within the very same second-level concept. If one wanted to use the word "point" in each of these geometries, it would become equivocal. To avoid this, we should have to add the name of the geometry, e.g. "point of the *A*-geometry," "point of the *B*-geometry," etc. Something similar will hold for the words "straight line" and "plane." And from this point of view, the questions of the consistency of the axioms and of their independence from one another (that is, of the unprovability of certain propositions from certain presuppositions) will require reexamination. One could not simply say "the axiom of parallels," for the different geometries would have distinct axioms of parallels. If the wording of each of these were the same, this would mistakenly have been brought about by the fact that one had simply said, for example, "straight line" instead of "straight line of the *A*-geometry." This way of talking may veil the difference of the thought-contents, but it certainly cannot remove it.

But herewith we have already reached the beginning of a path that leads to greater depths. Perhaps I shall be allowed to pursue it at some future date.

On the Foundations of Geometry

BY A. KORSELT IN PLAUEN I. V.

This paper is intended as a contribution toward reaching an understanding concerning the foundations of geometry. My discussion will be in reference to Mr. Frege's views on this subject, which are expressed on pp. 319 ff. and 368 ff. of this volume.*

To questions of the form "What is that?" one frequently has to reply with the counterquestion "Under what conditions will my answer be acceptable to you?" For any answer must be couched in words, concerning whose reference the questioner can ask once again; and thus there would be no end of questions—unless we proceed like Hegel, who explains "nature" as "pure being in itself," or give a definition as long as that given by the Supreme Court for the word "train." One must therefore agree beforehand, which assertions will be regarded as comprehensible.

If we call a true but logically unprovable thought an "axiom" (or, following the newer mathematicians, a "basic fact"), then "definitions" (nominal definitions, impositions of names) are not axioms. A "pure" judgment (a true proposition in general)

From *Jahresbericht, 12* (1903), 402-07.
* These page numbers refer to the original pagination of Frege's 1903 articles, "On the Foundations of Geometry." Cf. pp. 22 ff. and 29 ff. above. [Trans.]

states a(n objective) circumstance that is independent of being thought. It holds true whether or not it is thought by someone; whether or not it is communicated by means of these or those particular signs. But it is precisely such relation to a consciousness that the imposition of a name contains, and it lies in the words "let us call," "is called," etc. Thus although a (nominal) definition does not produce understanding of a truth, nevertheless it does produce understanding of concepts in that it combines concepts that prior to this were in our consciousness but without any connection. It is with good reason, therefore, that a new sign is assigned to such a combination.

Why shouldn't axioms and theorems contain signs whose "references" have not previously been settled? This prompts the further question of what is "referred to" by the assertion that the reference of a sign is settled. Many things. Either that, should the case arise, this sign will be recognized as the same sign for the same object; or that one will always recognize all or at least certain simple propositions using the sign as being either true, false, or definitions. "Mt. Blanc is 4800 m. high" is understood by many who never in their life have seen either the mountain or a height of 4800 m. because they know how to connect this proposition in a certain way with assertions with which they are familiar. If we should finally be unable to agree on the "reference" of the expression, this would merely be an indication of the fact that one or the other of us must acquire more propositions about this sign or using this sign. "The sign has no reference" will therefore mean, "We are not acquainted with any propositions regulating the use of this sign in general or for a given domain."

According to this, of course, no "theorem" can contain an "unknown" sign, else it would not be a theorem; it could not be derived from *known* propositions with known signs. But if an "axiom" contains a /403/ hitherto unknown sign, then it is simply the case that this "axiom" is itself a rule, a proposition determining the use of this sign!

To be sure, if the axiom is intended as a description of *known* facts, perhaps of perceptions, then all names occurring in it must designate an experience, whether it be one of intuition or one of thought.

But modern mathematics, which more and more blends into exact logic, no longer designates by its axioms (basic assertions) certain empirical facts (apart from their being thought itself), but at best *indicates* them, just as in algebra a letter does not determine a number but merely indicates it. "Arithmeticized," or better, "rationalized," mathematics merely arranges its principles in such a way that certain known interpretations are not excluded. In this way, *one* sequence of formal inferences can sometimes be "interpreted" in *different* ways. The "domain of application" of these axioms extends exactly as far as objects of experience that can be combined in the way described by the axioms, can be assigned to them.

We must therefore distinguish between those formal theories ("purely formal systems") that can be related to other experiences and those for which as yet no such correlation is known. The "objectiveness" and, above all, the consistency of a purely formal system is always and necessarily demonstrated by exhibiting objects of which the basic assertions hold. This, however, already lies outside the realm of the *formal* and mediates a transition to perception. Therefore whoever wants a logical and extremely simple connection of the totality of what we know cannot from the very start require objectiveness from a purely formal system. Even if we then call the latter "an empty playing with words, signifying nothing" and the like, as a strictly lawlike connection of propositions, it has no further need of any special "dignity." A purely formal system is noteworthy as long as the propositions that might occur in it do not lead to contradiction; indeed, it remains interesting even as a contradictory formal system, as long as this contradiction becomes apparent only at the end of a long sequence of inferences about the objects of the formal system. Therefore certain concepts of a "purely formal system" may even show themselves to be "empty." But on the other hand, a formal system may be "applied" to a *given* domain only after we have assured ourselves of the validity of the principles for that domain.

Therefore I would not say, as does Mr. Frege, that definitions must not contradict one another; rather, I would say that those interpretations of the signs of a purely formal system that are assertions must not lead to contradictions with accepted propositions. Otherwise we have at best defined not the *desired* concepts but different ones. If we do not follow this precept, we shall easily arrive at something false, as the following example will show:

If by *a, b, c,* we understand simple constructs of projective geometry ("null," point, plane, straight line, space, etc.; cf. *Mathematische Annalen,* vol. *44,* 156), and by *ab* and *a + b* the intersection and the union of *a* and *b* respectively, then the distributive law

$$a (b + c) = ab + ac$$

does *not* hold. /404/ However, if over and above the familiar axioms of the domain one also admits the basic proposition

$$b \nleq (a \nleq ab)$$

—for which no interpretation is known in projective geometry but which acquires a *logical* sense when *a* and *b* are understood as truth-values of assertions, *ab* is understood as the True or the False (depending on whether or not *a* and *b* are both true), and $a \nleq b$ is understood as the True or the False (depending on whether or not the truth-value of *b* is at least equal to that of *a*), then the distributive law is derivable once again, as Peano proves in *Rivista di matematica* (1891, p. 183).

It is the aim of all sciences to become exceedingly simple, objective, purely formal systems. However, most of them en-

counter serious difficulties in finding appropriate basic assertions. Rational mechancis, for example, already encounters this
problem.

Surely one would not say, "The axioms of the group A define the concept a" if group A were not the totality of those
axioms that one wanted to combine. Instead, one would say,
"Axiom-group A *determines* a certain concept (a-concept, e.g.
point-concept) of the highest level; group A', which contains
A, determines an a-concept of a lower level; and axiom-group
$A^{(n)}$, which includes all previously considered axiom-groups, determines the a-concept of the lowest level." Each of these a-
concepts is subordinate to the ones formed prior to it. We have
the right to give old names to these concepts if, upon appropriate interpretation, some signs of the "purely formal system"
become true propositions about the concepts designated with
these old names. Only, these old names must be modified; for
example, we have to say, "a-concept of a point" instead of
"concept of a point."

In this way, *one* formal theory can arise out of the combination or "encapsulation" of *many*. Only in the most comprehen-
sive of these theories may the signs occurring in it receive the
previously known names without any addition. Mr. Hilbert meets
this requirement when he talks of Euclidean, non-Euclidean,
Pascalian, Legendrean, Archimedean, non-Archimedean, etc.
geometries, depending upon the nature of the axiom-groups.
These epithets, of course, also apply to the constructs of these
"geometries."

As long as a has the same or a similar wording in all geometries, it is harmless to talk of "*the* proposition a" (e.g. *the*
axiom of parallels).

It is irrelevant whether it is the axioms or the characteristics
of the concepts introduced that are said to be consistent. The
former corresponds more closely to ordinary usage, according
to which two propositions are called "independent" of one
another if under certain circumstances both, under other cir-

cumstances not both, obtain; whereas they are called "incompatible" if there are no conditions under which both are satisfied together.

The words "basic proposition," "axiom," "definition" have the same reference as far as purely *formal* systems are concerned, namely that of "direct objects" (propositions); opposed to these stand the "indirect objects" (propositions). This distinction holds because *every* formal theory presupposes that *general* formal theory, whose interpretation (possibly among others) is formal logic itself. Otherwise no "derivation," no "proof," would be possible.

As Mr. Frege quite correctly indicates, the concepts of Euclidean and projective geometry differ with respect to their contents. As regards their extension, however, /405/ they agree, since both formal theories contain the same propositions albeit in a different order. On the other hand, it is because of this that, despite many common properties, "Euclidean point" differs from "Riemannian point" both in content and in extension. In view of the exhaustive works of Peano (*Principii di geometria,* Torino, 1889), Pieri (*I principii della geometria di posizione, della geometria elementare, memorie dell'acc. reale di Torino, 48* and *49,* [1899-1900]), and Hilbert, we can no longer refuse the designation "purely formal system" to geometry.

If we understand the "determination of concepts" as above, we shall remain within accepted usage when we say: A principle *a* is said to be dependent upon principles *b, c* . . . if *a, b, c* . . . belong to the same formal theory and if *a* is only apparently a direct proposition, whereas in reality either *a* itself or its negation stands in the relation of consequence to the propositions *b, c* . . . Certain given axioms "have an objective sense" if they are components of a formal theory whose figures can be interpreted as propositions of an already *familiar* and accepted discipline; they have the "desired sense" if they and their consequences can be interpreted as the *totality* of known propositions of a given discipline. However, it may be the case that a given set of principles is "invalid" with respect to a given set of truths because together with the *general* logical principles, they lead to a denial of some of these truths. Therefore Mr. Hilbert's expression "invalid axioms" has its point.

The assertions that give a "sense" to axioms are not components of a formal theory. They are like the presentation of a picture of an object that is being talked about. Such "indications" are not premises of a theory, but merely help to understand it. The mistake of many geometrical proofs lies precisely in the fact that by means of such indications, they surreptitiously obtain new unstated principles. This is what made possible Kant's opinion that mathematics is "founded" not on concepts but on intuitions.

Modern mathematicians would not have fallen into contradictions or confusions ("For me, an axiom is that requirement by means of which I introduce precise assertions into imprecise intuition") if they had studied Bolzano's *Wissenschaftslehre* (Sulzbach, 1837). Bolzano, the great opponent of Kant, is the first philosophical mathematician and mathematical philosopher since Leibnitz. He is familiar to mathematicians only through his foundation of mathematics and as the discoverer of set-theory; his views concerning the foundations of geometry and concerning scientific theories in general (*Wissenschaftslehre,* §§ 79, 554-59), however, are at least as significant.

I believe that herewith I have dispelled the doubts contained in Mr. Frege's first note; those contained in the second seem to me to pose no greater difficulties. It is said that it is not at all obvious from Hilbert's "explanations," "to what the explained expressions refer." What exactly is wanted here? Perhaps an indication of a perception, a model, a pocket watch, or a written symbol that is supposed to bear the name that is to be explained, or that is supposed to satisfy the proposition to be explained? Such an "explanation" convinces us of the objectiveness of a purely formal system, and in the case of Mr. Hilbert's propositions we need not go far afield to find one. As for the rest, it does not belong to a *formal* theory but to its interpretations; or it is a means of impressing the theory upon our memory, of illustrating /406/ it. Such "explanations" must not be taken as *presuppositions* of the theory and consequently cannot influence it.

Or must the "explanation" and every principle be formulated in such a way that from it [alone] one can see whether or not it fits any object whatever, e.g. Mr. Frege's pocket watch? That, surely, no mere explanation can accomplish! For that, after all, one needs experience, i.e. perception! It would be as if one wanted to convey a picture of a straight line to someone who was congenitally blind, merely by means of a description. Whoever has any doubts as to whether or not his pocket watch satisfies a given proposition must simply refrain from applying this explanation to it until either he has learned decisive new propositions or a watchmaker has repaired his pocket watch. We humans, after all, are not omniscient; therefore patience! The fact that at present it is impossible to reach a decision concerning the "falling of an object under a concept" proves nothing against the correctness and utility of that concept. Or is our present explanation of "transcendental number" perhaps useless because it does not decide anything about the "Euler-Mascheron constant"?

Or perhaps every "explanation" is supposed to have this form:

$$a \text{ is a } b$$

or:

$$a \text{ is that which has the known constitution } b.$$

E.g. "A point is that which has no parts." "Unity is that according to which everything is called one." Such an explanation is useless, and consequently the Euclidean concept of a point is not a "concept of the first level," because neither it nor [for that matter] Newton's "explanation" (that the relation of any quantity to a quantity taken as a unit is called number) has

ever been a premise of a valid inference. Chr.[istian] Wolff, who
wanted to define all concepts nominally, of course failed for that
very reason. We simply have to begin our investigation with
some simple concepts that cannot be analyzed any further, otherwise there would be no end to our defining. The simple concepts
that are taken as basic can be determined only by propositions
in which such a concept occurs several times or several such
concepts occur simultaneously. Simple basic concepts do not
form a sequence but a net, in which we can indeed get from
each knot to every other but which for all that cannot be un-
raveled into a single strand. There can be no nominal definition
of a Euclidean point that is based solely on *concepts* and does
not relate to any *perceptions* whatever. For, together with the
remaining Euclidean axioms, it would have to determine this
"point" completely and unambiguously. However, the totality
of propositions of Euclidean geometry can also be interpreted
as propositions about sets of triplets of numbers. Therefore
there is no definition of a Euclidean point that fits only one
object or one previously given class of objects (in particular,
points of our visual space)—which is precisely what a nominal
definition intends to accomplish. And in general, in the case of
many *conceptual* complexes there are not merely one but several
empirical complexes that can be correlated to them; *one* conceptual complex may have *several* interpretations.

   It is both inexpedient and unfair to demand of a formal
theory that it give a *determinate* reference to the "figures"
(names)—e.g. "point"—which it has constructed on the model
of proper names or concept-names. On the contrary, a purely
formal concept must keep open for its "names" as wide a /407/
domain of interpretation as possible, even if a particular interpretation is intended, so as to create a simultaneous connection among as many experiences as possible. Propositions
having the same wording should, if possible, be proved only
once, even if they appear in different disciplines. Only in this
way can one achieve a synopsis of the completeness, purity,

and independence of the principles and basic concepts. The signs of a formal theory have no "reference" at all; its laws merely give rules to which the intended interpretations actually are subject and to which those not intended are supposed to be subject. The rules and propositions of any game, e.g. chess, are the best example for this. Many of Leibnitz's assertions show that he understood his *calculus ratiocinator* as a "formal theory" and that he intended to make even games subject to it. Surely this is sufficient to permit a philosopher to find a "thought" expressed even in a configuration of chess pieces! All propositions of a lawlike complex express thoughts, no matter what their object may be.

The above example of the Euclidean point shows that we cannot, in concert with Mr. Frege, demand that every system of principles be solvable for the unknowns (basic concepts) that occur in them; we certainly cannot demand an unequivocal solution. That would amount to unlimited defining.

Frege's example from number-theory proves nothing against the explanation of Hilbert which he was criticizing. After all, from these rules [alone] nothing is inferred in number-theory about the behavior of 2 and 8 modulo 3. It is only in conjunction with those propositions by means of which the concepts "two," "three," "eight," "sum," "number" are determined that something is inferred. And Hilbert's explanation, too, is intended to serve such a purpose. From a proposition *by itself* one naturally can only get exactly what was expressly put into it; else one will be guilty of a sophism. We cannot infer from an *explanation* of the intersection of two circles, that the circles intersect if the line joining their centers is smaller than the sum of their radii; but we can infer this from this explanation together with other known propositions.

Nor is it the case that "the ontological proof for the existence of God is brilliantly vindicated" by the axiom "On every straight line there are at least two points." For the former is intended to *prove* existence; the axiom *presupposes* it for all or at least some of the propositions that follow from it. In fact, the "existence-propositions" of exact logic and mathematics are no more than presuppositions of certain conditional propositions in whose "assertions" certain concepts mentioned in the existence-propositions no longer occur. Every proposition b utilizing the existence-proposition a_x could read like this:

$$a_x \nleq b, \text{ if } a_x, \text{ then } b$$

where the letter x can be replaced by any other letter of the alphabet but does not occur in b at all. If the signs for the basic existence-propositions that are used in the derivation of b are presented together with b, clarity has been sufficiently served, nor has rigor been impaired.

For these reasons I cannot, despite much thought about the matter, find that Mr. Frege's objections against Hilbert's presentation are justified.

On the Foundations of Geometry

BY GOTTLOB FREGE IN JENA

I am only too pleased to enter into a direct interchange with men who have directed their thoughts to the same questions as I. And therefore I at first welcomed Mr. Korselt's essay "On the Foundations of Geometry" (this journal, *12*, 402). And I also had the pleasure of discovering some points of contact between us. Let me first emphasize these. Mr. Korselt uses my expressions "truth-value," "the True," "the False"— and, as it seems, in my sense. However I should like to ask that he say "the truth-value *a*" instead of "the truth-value of *a*". The explanation which Mr. Korselt gives the sign-complex '*a* ≠ *b*' almost agrees with my explanation of the corresponding sign in my *Begriffsschrift*; and it is better than that of E. Schröder, who apparently introduced the sign, and than that of Peano for his '⊃'.

As for the rest, I have of course been disappointed by Korselt's criticism. It does not offer as suitable a foundation for understanding and fruitful development as I had wished.

If Mr. Korselt wanted to prove that my reservations concerning Hilbert's presentation are unjustified, then he should have examined all of my objections. This he did not do. And yet, a single irrefutable objection can bring the whole theory to ruin.

Mr. Hilbert's theorems concern axioms, their independence

From *Jahresbericht, 15* (1906), 293-309, 377-403, 423-30.

and their consistency. It is therefore imperative to leave no doubt concerning the sense of the word "axiom." And consequently it is a fault of Mr. Hilbert's paper, that it leaves this concept nebulous. I have been at pains to draw sharp boundaries; Mr. Korselt, it seems, diligently blurs them once again. How is this to be explained? Perhaps by a drive for self-preservation on the part of Hilbert's doctrine, for which an obscuring of the issue may /294/ well be a condition of survival. If this is correct, then a savior of the doctrine must of course seek to prevent clarification. I do not believe that Mr. Korselt has done this on purpose; but by his procedure he has made it inordinately difficult for me to reply to his essay. Mr. Hilbert, so far as I know, does not reply to my arguments at all. Perhaps in him too there is at work a secret fear, deeply shrouded in darkness, that his edifice might be endangered by closer investigation of my arguments. On the surface, of course, there will probably float the opinion that my arguments are simply not worthy of closer consideration. However, if Mr. Hilbert should ever come to illuminate the nether regions with the light of his understanding, he will perhaps allow the point that ultimately what is false in his doctrine cannot be maintained, but that it is useful and honorable to repudiate it, so that what is true and valuable may stand out all the more clearly and indisputably.

In my opinion, it was the task of a critic to take a stand *vis-à-vis* my premises; in which case the question would have been attacked at its roots and everything that followed would have been sharply illuminated. In my first essay on the present subject I considered it necessary to talk at some length about axioms and definitions, since it appeared to me that Hilbert's Festschrift threatened a great confusion. Mr. Korselt should first have addressed himself to my remarks on this point, since they constitute my point of departure and everything that follows is closely connected with them. To definitions that stipulate something, I opposed principles and theorems that assert something. The former contain a sign (word, expression) that is still to receive a reference by means of them; the latter contain no such sign. Does Mr. Korselt not admit the essential difference between these two types of propositions? Why not? He continues to talk of axioms that define something, as if this offered no difficulties at all. But what is an axiom really supposed to do? Is it supposed to assert something, or is it supposed to stipulate something? Why do we have two words, "axiom" and "definition," if axioms too are supposed to define? Merely in the interest of greater obscurity?

In the second edition of Mr. Hilbert's work as well, axioms merrily go on defining as though nothing had happened. Evidently Mr. Hilbert himself does not know what he means by the word "axiom"; and consequently it also becomes quite doubtful whether he knows what thoughts he connects with his propositions; and still more doubtful whether Mr. Korselt knows this. Or do these gentlemen perhaps consider the thought-content /295/ of propositions to be superfluous? Words! Words! Words! I have indicated a confusion of usage in Mr. Hilbert's Festschrift with respect to the word "axiom." Does Mr. Korselt admit this conflict, or does he believe that he can smooth it over? He ought to have talked about this clearly, at the very beginning.

At the beginning of my second essay, I draw attention to the fact that Mr. Hilbert's explanations and definitions appear to be of two kinds. As examples of the first kind, I adduce the explanation for points of a straight line lying on the same side of a point, and for points of a straight line lying on different sides of a point. Gauss's definition of number-congruence also belongs to this kind. As an example of an explanation of the second kind, I adduce that of the word "between." I construct another definition of number-congruence on the model of the latter. Now if Mr. Korselt wanted to dispute in a constructive manner, he ought to have addressed himself to this. Does he admit that the explanations and definitions of Mr. Hilbert are of two kinds? What justifies the choice of the same word, "axiom" or "explanation," for both? Does Mr. Korselt see the enormous difference between the Gaussian definition of number-congruence and the one which I constructed on the Hilbertian model? This ought to have been made clear; otherwise we lack any solid basis for a fruitful discussion.

I believe that with my exposition about the use of the words "axiom" and "definition" I move within the bounds of traditional usage, and that I may justifiably demand that one not cause confusion by a completely new usage. Still, I am willing to consider a completely new usage of these words; only I must demand that it be uniform, and that it be presented intelligibly and unequivocally, prior to any consideration of particular questions. This I miss in Mr. Korselt's exposition.

Nevertheless, I do accord a value to his treatise, which makes it appear advisable to consider it more closely. For it seems to me that Mr. Korselt wants to give a peculiar turn to Mr. Hilbert's doctrine in that he understands it as a formal theory, as a purely formal system. Whether this explication quite corresponds to Mr. Hilbert's intentions is another question; for all that, a lot speaks in favor of it.

In turning to a closer examination of Korselt's treatise, I shall first collect what we find scattered throughout it concerning axiom, definition, and reference.

/296/ There we read the proposition,

> If we call a true but logically unprovable thought an axiom, then "definitions" (nominal definitions, impositions of names) are not axioms.

What Mr. Korselt presents here in the antecedent clause as the meaning of the word "axiom" may no doubt be called the traditional, Euclidean meaning. Axioms differ from theorems in that they are unprovable. The reasons given by Mr. Korselt for his contention approximate those given by me and also show that conversely, no axiom can define anything. If the word "axiom" is taken in this sense, then the expression of an axiom must contain no unknown sign, for otherwise it would express no thought at all. Letters intended to lend generality to the content of a principle may of course occur in it, since it is known how they contribute to the expression of the thought even though they designate nothing. Herewith we have an answer to Korselt's question, "Why shouldn't axioms contain signs whose references have not previously been settled?"

Initially, it appears as though Mr. Korselt accepts this Euclidean meaning, and indeed nowhere does he explicity reject it. But almost all of his later statements stand in direct contradiction to it. Thus, he recognizes, for example, the possibility of invalid axioms. If he wanted to write clearly, then he ought to have said something like this: "However, I do not accept this traditional sense of the word 'axiom', but rather use it in the following sense." And then he ought to have presented the latter as clearly as possible.

In contradiction to the Euclidean meaning, Mr. Korselt assumes that axioms contain hitherto unknown signs and that they define or determine a concept. But the fundamental difference between axioms and definitions, which he himself initially emphasizes, makes it quite impossible that axioms define anything. One must be clear on this point: Is a principle supposed to stipulate something, or is it supposed to express a thought and assert it to be true? And whatever one decides here, one must adhere to it and not continually waver back and forth between different conceptions.

Mr. Korselt also has the unfortunate idea of blending Heine's formal theory of arithmetic with Hilbert's doctrine, and of understanding an axiom as a rule for the use of the signs occurring in it. This is a third conception of axioms.

/297/ In saying that modern[1] mathematics no longer designates certain facts of experience with its axioms but at best indicates them, Mr. Korselt brings the axioms of modern mathematics into contrast with those of Euclid; and doubtless we may

---

1. Are we still not out of this ghastly "modern" era?

assume that he counts himself among the modern mathematicians. Clearly, he also counts Mr. Hilbert among them and believes that with this proposition he has hit upon the latter's usage of the word "axiom". If this is correct, then it is a gross error to assume that Mr. Hilbert has shown anything at all about the dependence or independence of the Euclidean axioms; or that when he talks about the axiom of parallels it is the Euclidean axiom. Accordingly, by an axiom we apparently are to understand something which looks like a proposition that is supposed to express a thought, but which actually isn't a real proposition at all since it does not express a thought but merely indicates it; just as for example the letter 'a' in arithmetic is not a number-sign, since it does not designate a number but merely indicates one. This is a fourth conception of axioms. But it is not the last, for according to Mr. Korselt, axioms also describe the manner in which objects of experience can be combined. But how these conceptions are to be reconciled with one another, in what relations they might perhaps stand to one another—about this there hovers a mysterious darkness. And now consider that everything hinges on understanding the word "axiom," if one wants to understand Hilbert's propositions concerning the independence of these axioms. Clarity! Clarity! Clarity! Does Mr. Korselt think that it is a pleasure to be led by him through these thickets?

We can probably characterize the situation by means of the following analogy: Mr. Hilbert chops both definitions and axioms very fine, carefully blends them, and makes a sausage out of this. Mr. Korselt is not satisfied with this mixture. He also minces Thomae's rules concerning the use of signs, adds a pinch of my indicating, from his own resources adds a description of the manner in which objects of experience can be combined, mixes it all up very thoroughly, and makes a sausage out of this. At least there is no dearth of ingredients; and I have no doubt but that something good will result for the fancier of such delicacies.

Let us take an example!

Every anej bazet at least two ellah.

/298/ "How could anyone write such hair-raising nonsense! What is an anej? What is an ellah?" So I hear it being asked with indignation. At your service! That is an axiom, not of the Euclidean, but of the modern, kind. It defines the concept *anej*. What an anej might be is a very impertinent question. We should first have to discuss under what circumstances an answer would be acceptable. If we don't find a thought in this axiom, that matters little. The proposition does not claim to be a description of known facts; at best it indicates them, and that, moreover, only very delicately. For example the well-known empirical fact that every sausage has at least two ends, or that every child waved at least two pennons—It is obviously the description of a way in which objects of experience—the pennons—can be combined with one another. The domain of application of this axiom extends precisely as far as objects of experience can be assigned to the axiom.

Mr. Korselt discusses the word "reference" by attacking my thesis that axiomatic propositions must contain no proper names, concept-words, or relation-words whose reference has not previously been settled. I here use the word "axiom" in the old Euclidean sense; Mr. Korselt, however, uses it in a modern one. Consequently there is not the slightest conflict between us, since what Mr. Korselt denies is completely different from what I maintain. But with this, another divergence becomes apparent. I had thought the matter to be much simpler, namely thus: In scientific use, a proper name has the purpose of designating an object; and in case this purpose is achieved, this object is the reference of the proper name. The same thing holds for concept-signs, relation-signs, and function-signs. They designate concepts, relations, and functions respectively, and what they designate then is their reference. According to Mr. Korselt, however, the matter is not quite so simple. According to him, the assertion that the reference of a sign is settled means either that should the sign occur, it will be recognized as the same sign for the same object; or that one will always recognize all or at least certain simple propositions using the sign as being true, false, or definitions. Let us ask whether, according to this, the reference of the word "anej" is settled. At the present time, only one proposition using the word is known, namely the axiom adduced above. Is the latter, then, true, false, or a definition? True? That does not seem to be quite right. False? Equally as little. But a definition! That's it: it is a defining axiom, and I dare say we may be confident that we shall always recognize it again as such. Therefore the reference of the word "anej" is settled.

/299/ According to Mr. Korselt, the words "principle," "axiom," and "definition" have the same reference as far as a purely formal system is concerned, namely that of direct objects (propositions); and opposed to these stand the indirect objects (propositions). This illuminates the close relationship which according to his usage obtains between axioms and definitions. A difference is not mentioned. As with Mr. Hilbert, the definition probably consists of axioms, whereby it is not ruled out that a definition may contain only a single axiom.

We get to know little about what constitution a proposition must have in order to be an axiom, a definition, or a rule about the use of signs. Actually, we get to know no more than that it must contain unknown signs. Of course there is still this puzzling assertion:

"Those interpretations of the signs of a purely formal system that are assertions, must not lead to contradictions with accepted propositions. Otherwise we have at best defined not the *desired* concepts, but different ones."

I really find all of this quite unintelligible. At first it looks as if the above were intended to limit the interpretations; in the end it looks like a rule for giving definitions. I don't quite see how the fact that the defined concept is one we wanted can prevent a contradiction that would emerge if the concept were not one we wanted. Even the example does not clarify the matter for me. What is here the definition? Where is the desired concept? What interpretation do we have? With what accepted proposition does a contradiction ensue? Can the latter not be avoided by a different interpretation? Does the mistake lie in the interpretation or in the definition?

With respect to defining, Mr. Korselt detects a difficulty. Mr. Korselt says that the answer to the question of what something is must always be given in words, concerning whose reference we can ask once again; and thus there would be no end of asking. And similarly, he further states, "We simply have to begin our investigation with some simple concepts that cannot be analyzed any further." Very well! But what is this supposed to prove? To begin with, it surely shows no more than that it is unreasonable to want to define everything. But it almost looks as if the question concerning reference were thereby to be denied justification altogether, and as if defining, or at least a particular manner of defining, were to be presented as worthless. It cannot be allowed that only a particular manner of defining would be affected by this; for no matter of what type a definition may be, it will always have to /300/ presuppose certain words or signs as known. Initially one may think that Mr. Korselt wants to stop defining with simple concepts that cannot be analyzed any further, since otherwise there would be no end of defining. But he continues: "The simple concepts that are taken as basic can be determined only by propositions in which such a concept occurs several times or several such concepts occur simultaneously."

Simple concepts, then, are to be determined as well. And is this not a definition? In which case is there any end to defining? Later, of course, it emerges that such propositions do not determine the concepts: "It is both inexpedient and unfair to demand of a formal theory that it give a *determinate* reference to the 'figures' (names)—e.g., 'point'—which it has constructed on the model of proper names or concept-names."

Mr. Korselt says, "The above example of the Euclidean point shows that we cannot, in concert with Mr. Frege, demand that every system of principles be solvable for the unknowns (basic concepts) that occur in them; we certainly cannot demand an unambiguous solution. That would amount to unlimited defining."

The danger of having to define *ad infinitum* arises if and only if one demands that everything be defined. But who forces us to do this? On the other hand, this danger has nothing to do with the demand that a system of principles be capable of unambiguous solution. My demand holds only in that case where such a system is supposed to be a definition by which certain concepts or relations are to be determined. Without the possibility of an unambiguous solution, we simply do not have a determination. My demand stems from the very nature of the matter; and if it cannot be met, then it does not follow that it must be dropped. Rather, it follows that concepts and relations cannot be determined in this manner; and that therefore Mr. Hilbert's so-called definitions, which are of this kind, do not meet the requirements we must place on definitions. When Mr. Korselt proves that in the case of the definition of a point, the requirements a definition must meet cannot be met—I do not want to embark on an examination of this proof—then from this it follows that we should abstain from defining; not, however, that we should do as we please and call the result a definition of a point. If it has been proven that a leap across a given ditch is impossible, then from this it follows that we had better not attempt it; not, however, that we should limp and call this limping /301/ a leap across the ditch. Whoever willfully deviates from the traditional sense of a word and does not indicate in what sense he wants to use it, whoever suddenly begins to call red what otherwise is called green, should not be astonished if he causes confusion. And if this occurs deliberately in science, it is a sin against science.

What is actually the purpose and nature of Hilbert's defining? And to what laws is it subject? For surely not even Mr. Korselt himself believes that every system of propositions that contain certain unknown words like "anej" and "bazen" constitutes an explanation of these words. We are agreed in this, that an explanation or definition generally need not point to a perception—a model. However, we must come to an understanding not only about what an explanation does not need to achieve, but also, and mainly, about what we must require of it.

My opinion is this: We must admit logically primitive elements that are indefinable. Even here there seems to be a need to make sure that we designate the same thing by the same sign (word). Once the investigators have come to an understanding about the primitive elements and their designations, agreement about what is logically composite is easily reached by means of definition. Since definitions are not possible for primitive elements, something else must enter in. I call it explication. It is this, therefore, that serves the purpose of mutual understanding among investigators, as well as of the communication of the science to others. We may relegate it to a propaedeutic. It has no place in the system of a science; in the latter, no conclusions are based on it. Someone who pursued research only by himself would not need it. The purpose of explications is a pragmatic one; and once it is achieved, we must be satisfied with them. And here we must be able to count on a little goodwill and cooperative understanding, even guessing; for frequently we cannot do without a figurative mode of expression. But for all that, we can demand from the originator of an explication that he himself know for certain what he means; that he remain in agreement with himself; and that he be ready to complete and emend his explication whenever, given even the best of intentions, the possibility of a misunderstanding arises.

Since mutual cooperation in a science is impossible without mutual understanding of the investigators, we must have confidence that such an understanding can be reached through explication, although theoretically the contrary is not excluded.

/302/ Are, then, Hilbert's definitions explications? Explications will generally be propositions that contain the expression in question, perhaps even several such expressions; and herein they agree with what Mr. Korselt states with the following words: "The simple concepts that are taken as basic can be determined only through propositions in which such a concept occurs several times or several such concepts occur simultaneously." If Hilbert's definitions were to serve only the mutual understanding of the investigators and the communication of the science, not its construction, then they could be considered explications in the sense noted above and could be accorded all the consideration to which as such they could lay claim. But they are intended to be more. It is not intended that they belong to the propaedeutic but rather that they serve as cornerstones of the science: as premises of inferences. And given these demands, they cannot be accorded the leniency of judgment which they could have demanded as mere explications. Moreover, even as explications they miss their mark: namely to make sure that all who use them henceforth also associate the same sense with the explicated words. We are easily misled by the fact that the words "point," "straight line," etc. have already been in use for a long time. But just imagine the old words completely replaced by new ones especially invented for this purpose, so that no sense is as yet associated with them. And now ask whether everyone would understand the Hilbertian axioms and definitions in this form. It would amount to pure guesswork. Some would perhaps not be able to guess anything at all; some this, others that.

Let us turn to proper definitions! They, too, serve mutual understanding, but they achieve it in a much more perfect manner than the explications in that they leave nothing to guesswork; nor need they count on cooperative understanding and goodwill. Of course they do presuppose knowledge of certain primitive elements and their signs. A definition correctly combines a group of these signs in such a way that the reference of this group is determined by the references of the signs used. From a purely theoretical point of view, this might suffice; but such sign-groups often become too unwieldy and are too time-consuming to utter or to write out. We need a simple sign for them. And it is the task of the definition to give this new sign to the content determined by the familiar signs. Now it may happen that this sign (word) is not altogether new, but has already been used in ordinary discourse or in a scientific treatment that precedes the truly systematic one. /303/ As a rule, this usage is too vacillating for pure science. But if we assume that in a given case it satisfies the most stringent demands, then one might think that in that case a definition would be unnecessary. And if, like an explication, a definition were to serve only mutual understanding and the communication of the science, then in this case it would indeed be superfluous. But that is an advantage gained only incidentally. The real importance of a definition lies in its logical construction out of primitive elements. And for that reason we should not do without it, not even in a case like this. The insight it permits into the logical structure is not only valuable in itself, but also is a condition for insight into the logical linkage of truths. A definition is a constituent of the system of a science. As soon as the stipulation it makes is accepted, the explained sign becomes known and the proposition explaining it becomes an assertion. The self-evident truth it contains will now appear in the system as a premise of inferences.

The mental activities leading to the formulation of a definition may be of two kinds: analytic or synthetic. This is similar to the activities of the chemist, who either analyzes a given substance into its elements or lets given elements combine to form a new substance. In both cases, we come to know the composition of a substance. So here, too, we can achieve something new through logical construction and can stipulate a sign for it.

But the mental work preceding the formulation of a definition does not appear in the systematic structure of mathematics; only its result, the definition, does. Thus it is all the same for the system of mathematics, whether the preceding activity was of an analytic or a synthetic kind; whether the definiendum had already somehow been given before, or whether it was newly derived. For in the system, no sign (word) appears prior to the definition that introduces it. Therefore so far as the system is concerned, every definition is the giving of a name, regardless of the manner in which we arrived at it.

It is self-evident that what is given a name (sign) must be determined by the definition. A word without a determinate reference has no reference so far as mathematics is concerned.

Now in my second essay I have shown that for the most part, Mr. Hilbert's definitions miss their mark; at least they do so if we assume that they are to assign references to the words "point," "straight line," "between." /304/ Mr. Korselt is basically of my opinion, for he says, "The signs of a formal theory have no reference at all."

Now clearly, he takes Mr. Hilbert's theory to be a formal one. Accordingly, there can occur in it no explanations of signs such that by means of these the signs are given references. Therefore if—as I claim—there is no concept that is assigned as reference to the word "point" by Hilbert's definition, then this is quite in accord with the fact that according to Mr. Korselt, in a formal theory no reference whatever corresponds to the word "point," and that by the very nature of a formal theory none could correspond to it. In agreement with this, Mr. Korselt states, "It is both inexpedient and unfair to demand of a formal theory that it give a *determinate* reference to the 'figures' (names)— e.g. 'point'—which it has constructed on the model of proper names or concept-names."

What is asserted here seems to be a bit weaker than what was said at the place mentioned previously; but as a matter of fact, it is one and the same, for a sign without determinate reference is a sign without reference. Therefore complete agreement obtains! But then, why does Mr. Korselt quarrel with me, e.g. over my example of the pocket watch? If the word "point" does not designate a concept for Mr. Hilbert, then it goes without saying that the question of whether my pocket watch is a point cannot be answered. Mr. Korselt could simply have pointed out that so far as the formal theory is concerned, it isn't even the purpose of the word "point" to refer to something. Then of course he would have had to say what the purpose of Hilbert's explanations and definitions really is, since it cannot be the one of giving references to signs. It may surely be taken as established once and for all that Hilbert's pseudo-definitions do not give references to the words "point," "straight line," "between," etc. that seem to be explained by them. And herewith we have achieved one of the aims I have pursued. While agreeing with me in this conclusion, Mr. Korselt appears to raise objections against the way in which I have reached it. This may seem unimportant; nevertheless, it will not be superfluous to examine the matter a little more closely.

I demand from a definition of a point that by means of it we be able to judge of any object whatever—e.g. my pocket watch —whether it is a point. Mr. Korselt, however, misunderstands this to mean that I demand that the question be decidable from the definition alone, without the /305/ help of perceptions; and he maintains that this is impossible. Quite right! The question of whether a given stone is a diamond cannot be answered by the mere explanation of the word "diamond" itself. But we can demand of the explanation that it settle the question objectively, so that by means of it everyone well acquainted with the stone in question will be able to determine whether or not it is a diamond. Therefore if it is merely on account of our incomplete knowledge of the object that we cannot answer the question, then the explanation is not to blame. If, however, the question must remain unanswered no matter how complete our knowledge, then the explanation is faulty. And this is presently the case. The very same difficulty that arises with the question of whether my pocket watch is a point arises for every object, whether it be sensibly apprehensible or not. For example, if we take the number two or even a Euclidean point, we shall always encounter the same obstacle, namely that we shall already have

to know of another object that it is a point before we could
even begin to decide whether the statement of Hilbert's first
axiom applies. And no knowledge of the presented object, be it
ever so complete, can help us overcome this difficulty. Therefore even if I agree with Mr. Korselt that the mere definition
does not suffice to answer the question whether an object falls
under the defined concept but that a sufficient knowledge of the
object, acquired somewhere, is also necessary, I must nevertheless insist that a complete knowledge of the object together with
the definition must suffice. A repair of the watch does not help
in the present case, for the fault does not lie in it but in Hilbert's
definition, which provides neither sense nor reference for the
grammatical predicate "is a point," so that every proposition
with this predicate is senseless, no matter what its subject may be.

The situation is exactly the same with the example from
number-theory. The Gaussian definition of number-congruence
fulfills its purpose because it reduces the reference of the word
"congruent" to the references of the expressions "different" and
"a number divides evenly into a number," which are already
known: It constructs the reference logically out of known building blocks. What I miss in Hilbert's definitions, for example in
that of the word "between," is by no means the indication of a
model—of the perception—but rather the logical construction.
I hear the reply, "A reduction to what is known is not possible
here." Well, in that case no definition is possible. In that case
we shall simply have to recognize a primitive element and be
satisfied with an elucidation. The latter, however, /306/ cannot
appear in the system but rather must precede it. Within the
system, one will just have to presuppose the word "between"
as known; just as one can never circumvent the necessity of
assuming some words to be known.

If a relation is correctly defined, then this definition together
with an adequate knowledge of any given object must suffice to
decide whether these objects stand in the defined relation to each
other. This knowledge is naturally expressed in propositions

that do not presuppose an acquaintance with the relation-sign in question, and that therefore contain neither it nor an expression that would have to be explained in terms of it. I have never demanded that the decision be possible on the basis of the definition alone, without the aid of other propositions. If Mr. Korselt wants to refute me, he has only to deduce from my definition of number-congruence which I constructed on Hilbert's model, that 2 is congruent to 8 modulo 3; and here he may use any propositions of arithmetic for whose understanding the word "congruent" is not necessary. Let him try it, and he will see that it does not work. Mr. Korselt may take whatever propositions about points on a straight line he pleases, as long as these do not presuppose an acquaintance with the word "between": using these propositions, he will still be unable to prove from Hilbert's definition of lying-between that one or the other of three points on a straight line lies between the other two.

However, if we posit the Gaussian definition of number-congruence, then in order to recognize that 2 is congruent to 8 modulo 3, we need only the propositions

'$8 - 2 = 3 + 3$' and '3 goes evenly into $3 + 3$'

which neither contain the sign for congruence nor presuppose knowledge of it.

We saw that a definition which is to assign a reference to a word must determine this reference. Hilbert's pseudo-definition does not achieve this. There is no relation that would be designated by the word "between" in accordance with this pseudo-definition. And here we find ourselves once more in complete agreement with Mr. Korselt, who does not want to accord any reference whatever to this word in a formal theory. Or might it perhaps have different references?

Herewith we come to a discussion of the requirement of unambiguous signs which, it seems, is ignored by some newer mathematicians—in contradistinction to the venerable Goethe, who, although /307/ not a mathematician, was nevertheless no fool. In the ninth book of his *Fiction and Truth* he says, "Just as, after all, anything can be asserted if we permit ourselves to use and apply words quite indeterminately, now in a wider, now in a narrower, in a more closely or a more distantly related sense."

Indeed, if it were a matter of deceiving oneself and others, there would be no better means than ambiguous signs.

Mr. Korselt declares that it is not risky to talk of "the theorem *a*" (e.g. the axiom of parallels) if *a* has an identical or similar wording in all geometries, as if the sense did not matter at all. I had given grounds to the contrary; Mr. Korselt ignores them and simply opposes his authority to them. Does this suffice? Well, his manner of talking is indeed practical if it is a question of imagining that what one has proved of axioms taken in the modern sense also holds of them when taken in the Euclidean sense.

Generally speaking, the sciences, too, are really moving in the opposite direction. They seek to make their language ever more precise by formulating technical expressions with the greatest possible precision, so as to escape the vacillations of ordinary usage. It is only the modern mathematicians who sometimes seem to seek their strength in ambiguity; and indeed it is decidedly convenient, inviting to flights of fancy. And yet the dangers arising for the certitude of the proofs associated with this ambiguity, cannot be ignored. Are they to be deemed nothing, then? One would think that the friends of ambiguity of signs would first indicate and fully justify precautionary rules that could safeguard against the dangers arising from this ambiguity. To my knowledge this has not occurred, and indeed would be a useless labor; for the appearance that ambiguous signs are necessary arises from unclear thinking and insufficient logical insight.

If one wants to defend the ambiguity of signs, one may in the first instance think of the use of letters in mathematics. But these letters are of a nature completely different from that of the number-signs '2', '3', etc., or the relation signs '=', '>'. They are not at all intended to designate numbers, concepts, relations, or some function or other; rather, they are intended only to indicate so as to lend generality of content to the propositions in which they occur. Thus it is only in the context of a proposition that they have a certain task to fulfill, that they are to contribute to the expression of the thought. But outside of this context, they say nothing. /308/ It is quite wrong to think that the proposition $'(a + b) \times c = a \times c + b \times c'$ expresses different thoughts, among others also the one contained in the proposition $'(2 + 3) \times 7 = 2 \times 7 + 3 \times 7'$. Rather, the first proposition expresses only a single thought which, however, is different from that of the second proposition. It is equally as wrong to think that the letter a now designates the number 2, now another number, or even several numbers at once. It simply does not have the purpose of designating a number, as does a number-sign; or for that matter, of designating anything at all. Rather, it has the sole purpose of lending generality of content to the proposition $'(a + b) \times c = a \times c + b \times c'$, and it is precisely this generality that differentiates this proposition from the second one.

Concept-words offer another occasion where it may seem that ambiguous signs are necessary. If we think that the word "planet" designates at one time the Earth, at another Jupiter, then we should take it to be ambiguous. But in fact it does not stand to the Earth in the relation of sign to thing signified. Rather, it designates a concept, and the Earth falls under it. No ambiguity is to be found here. Let us suppose that the word "planet" is unknown and that we wanted to designate the appropriate concept. We might then perhaps hit upon the idea of using the proper name "Mars" for it, and might find unreasonable the demand that the word "Mars" be given a determinate reference: as wide a range of interpretations as possible ought to be kept open for this name. But as a concept-word, "Mars" would have to be just as unambiguous as it would have been as a proper name. Do not say that as a concept-word it has no determinate reference, or that it refers to an indeterminate object. Every object is determinate; "indeterminate object" is contradictory, and wherever this expression occurs, we can be quite certain that a concept is what is really meant. We cannot say that the proposition '$x > 0$' assigns an indeterminate object, an indeterminate number, to the letter 'x' as its reference. Rather, what is designated here is a concept: *positive number;* nor is 'x' introduced as a sign for this concept; it merely takes the place of the proper names (number-signs) of objects that may perhaps be subsumed under the concept. Thus the appearance of ambiguity arises only out of an insufficient understanding, in that proper names and concept-words are not distinguished sharply enough.

A similar thing may occur one level higher, when talking of an indeterminate concept or the indeterminate reference of a concept-word, where what one really has in mind is a concept of the second level. /309/ This is probably the case when Mr. Korselt finds it inexpedient and unreasonable to give a determinate reference to the word "point"; and when he wants to leave open to names as wide a range of interpretations as possible. He apparently does not mean by this that a concept with as wide an extension as possible is to be correlated with the word "point"; for the latter still would be—indeed, would have to be—completely determinate. Rather, he has in mind a second-level concept within which, aside from the Euclidean *point-concept,* still other concepts fall. Of course this second-level concept must also be a completely determinate one; but it behaves toward the first-level concepts falling within it in a way similar to that in which a first-level concept behaves toward the objects falling under it. When we consider the multiplicity of these concepts of the first level (point-concepts), we get the notion that we are faced with an indeterminacy or ambiguity. This need not be the case here any more than it is in the case of the first-level concept *prime number,* where not only 2 but also 3 falls under the latter.

In no way is it necessary to have ambiguous signs, and consequently such ambiguity is quite unacceptable. What can be proved only by means of ambiguous signs cannot be proved at all.

II.

/377/ Before discussing the turn which Mr. Korselt gives to Hilbert's doctrine by calling it a formal theory or a purely formal system, I should like to adduce some considerations whose upshot is important for the understanding and appraisal of this very turn.

Mr. Korselt does not always appear to distinguish a proposition as what is sensibly perceptible, from the thought which is its sense. What I call a proposition *tout court* or a real proposition is a group of signs that expresses a thought; however, whatever has only the grammatical form of a proposition I call a pseudo-proposition. Examples of the latter are often to be found as antecedent and consequent propositions of conditional propositional complexes. We frequently encounter the conception of a conditional judgment as something by means of which judgments (propositions) are brought into relation with each other. But this only rarely applies, even if we say "thought" instead of "judgment"; for it is frequently the case that we have a thought in neither the antecedent nor the consequent proposition in themselves, but only in the propositional complex as a whole. Let us consider the proposition "If something is greater than 1, then it is a positive number." "Something" and "it" refer to one another. If we break this connection by separating the propositions, each of them becomes senseless. "It is a positive number" says nothing. To be sure, we can find a thought expressed in the proposition "Something is greater than 1"; namely, that there is something which is greater than 1. But it is not in this sense that the grammatical proposition occurs as antecedent of the propositional complex. We can also express this thought by utilizing the letter 'a' as in arithmetic:

If $a > 1$, then $a > 0$.

Here the letter 'a' only indicates, as did the words "something" and "it" above. The generality extends to the content of the whole propositional complex, not to the antecedent proposition by itself nor the consequent proposition by itself. Since neither the former nor the latter by itself expresses a thought, neither of them is a real proposition. The whole propositional complex is one; it expresses a single thought /378/ which cannot be divided into component thoughts. We can also express this thought thus:

Whatever is greater than 1 is a positive number.

The first grammatical proposition actually takes the place of the subject, and the second contains the predicate belonging to it. From this it is also clear that logically speaking we have only a single proposition. Here we do not have a relation between thoughts, but the relation of subordination of the concept *greater than 1* under the concept *positive number*.

The thought of the proposition

If the square of something is 1, then its fourth power is also 1

we can also express like this:

If $a^2 = 1$, then $a^4 = 1$

or also as

Whatever is a square root of 1 is also a fourth root of 1

or also as

Every square root of 1 is also a fourth root of 1

Here the subordination of concepts is once more discernible, so that even here we have only a single thought. Here, too, the grammatical component-propositions are merely pseudo-propositions without thought-content. For the letter 'a' in '$a^2 = 2$' refers to 'a' in '$a^4 = 1$' just as, for example, in Latin the word *quot* occurring in an antecedent refers to *tot* in the appropriate consequent. And just as the separation of such propositions renders both of them senseless, so in this case as well, what is contributed by the letter 'a' to the expression of the thought is lost by the dissolution of the propositional complex. The letter is supposed to lend generality of content to the whole proposition, not to the component pseudo-propositions. And thus it comes about that the whole propositional complex expresses a true thought, even though it contains a letter that signifies nothing; while the component pseudo-propositions have no sense because they contain the letter 'a', which neither has a sense nor lends generality of content to even one of these components. If it were supposed to do the latter, then '$a^2 = 1$' would indeed express a thought, albeit a false one; to wit, that every object is a square root of 1. But it is not in this sense that '$a^2 = 1$' occurs as a part of the thought-complex. From this we see that a proposition can express a true thought even though it contains words ('something', 'it') or letters that do not refer to anything /379/ but merely indicate whenever these words or letters have the purpose of lending generality of content to the proposition. On the other hand, we see that a grammatical component-proposition that contains such words or letters does not express a thought if the generality brought about by them is not supposed to be restricted to this component-proposition. In such a case, the grammatical component-proposition does not say anything; it is merely a pseudo-proposition. We can say neither that it is valid nor that it is invalid—insofar as we call a proposition invalid when it expresses a false thought.

To be sure, '$a^2 = 1$' contains something having a reference, and its reference is the concept *square root of 1*. However, what designates this concept is not the whole '$a^2 = 1$', but only that part of it which remains when we detach 'a'. Similarly for '$a^4 = 1$'.

In general, we may say that neither an antecedent nor a consequent pseudo-proposition expresses a thought or sense, although both are parts of a propositional complex that does express a thought, and though both may have components that have a sense.

Moreover, the use of the letter 'a' in such cases is fundamentally the same as that in the proposition

$$a^2 - 1 = (a - 1) \times (a + 1)$$

Here, too, it is supposed to lend generality of content to the proposition. The fact that in the first case we have two grammatical propositions whereas here we have only one is merely an inconsequential difference of form.

Insight into the logical nature of a mathematical theory is frequently made more difficult by the fact that what really ought to be represented as a unitary propositional complex is torn apart into apparently independent grammatical propositions. This often happens for stylistic reasons, in order to avoid a propositional monstrosity; but this cannot be permitted to obstruct one's insight into the nature of the case. For example, one begins like this: "Let a be . . ."—a locution in place of which the incorrect "Let a refer to . . ." is of course frequently preferred. Such propositions with different letters may in part precede the derivation, in part be inserted into it. Thus one finally arrives at a conclusion that is expressed in one proposition containing the letters that had apparently been explained previously; for propositions like "Let a refer to . . ." look like explanations that are supposed to give references to the letters. This appearance, however, vanishes upon closer examination. Let us take an example! The propositional complex "If a is a whole number, then (a $\times$ ($a - 1$)) is an even number" can be divided into two apparently independent propositions: /380/ "Let a be a whole number. $(a \times (a - 1))$ is an even number." But the first proposition cannot be considered an explanation of the letter 'a' such that this 'a' together with the reference thus acquired occurs in the second proposition; for this 'a' occurs in both propositions in place of a proper name. Therefore if it were to be given a reference, it could only be that of a proper name, i.e. an object. This, however, cannot come about by means of the proposition "Let a be a whole number," because this is not an identity-proposition but rather a subsumption-proposition. We cannot even say that although 'a' is not given a determinate reference, nevertheless it is given an indeterminate one—for an indeterminate reference is not a reference. There must be no ambiguous signs. The following consideration also shows that an antecedent pseudo-proposition should not be taken as an explanation of a letter occurring in it. For the proposition above can be brought into the form "If $(a \times (a - 1))$ is not an even number, then a is not a whole number." If we were to treat this proposition as before, we should end up considering the proposition "Let $(a \times (a - 1))$ not be an even number" as an explanation of the letter 'a'; and this would contradict our first explanation. Therefore we must not let ourselves be deceived by the fact that for stylistic reasons, an antecedent pseudo-proposition occasionally occurs in such a form that upon cursory examination it appears to be an explanation of one or more letters. For in fact neither these putative explanations nor the proposition in which the conclusion is asserted are real propositions. Rather, being antecedent and consequent pseudo-propositions, they belong inseparably together, so that only the whole constituted of them is a real proposition. It would greatly facilitate insight into logical structure if what is a single real proposition according to its subject matter were also a unitary propositional complex according to its grammar and did not break down into independent propositions. To be sure, in our word-languages such propositional complex would sometimes attain a monstrous length, whereas, because of its perspicuous

nature, the Begriffsschrift is better suited to the representation
of the logical fabric.

The use of letters is actually the same in all of these cases,
however different it may seem. They are always supposed to
lend generality of content to the whole, even when this whole
consists of apparently independent propositions. Of course,
words like "something" and "it" may also occur in place of these
letters.

A system of general theorems that coincide in their antecedent pseudo-propositions may be called a theory. Since the
consequent pseudo-propositions may be conjoined into a single
one by the use of "and," at least theoretically there exists the
possibility /381/ of changing the theory into a single theorem
consisting of antecedent pseudo-propositions and one—generally
composite—consequent pseudo-proposition. This theorem can
be given generality of content by letters or appropriate words.
The antecedent pseudo-propositions are what are sometimes
called presuppositions.

Now we can pass from the general to the particular by means
of an inference. So, for example, we can get from the proposition

$$a^2 - 1 = (a - 1) \times (a + 1)$$

to the proposition

$$5^2 - 1 = (5 - 1) \times (5 + 1);$$

and from the proposition

If $a^2 = 1$, then $a^4 = 1$

to the proposition

If $1^2 = 1$, then $1^4 = 1$

or even to the proposition

If $2^2 = 1$, then $2^4 = 1$.

As we can see, the external procedure here is to replace the letter that merely indicates by a referring sign. And similarly in other cases: in an inference from the general to the particular, the indicating letters or words are replaced by referring ones. General affirmative or negative propositions must first be brought into the conditional form. From the second example we see that the pseudo-propositions '$a^2 = 1$' and '$a^4 = 1$' thus yield the real propositions '$1^2 = 1$' and '$1^4 = 1$', or even the real propositions '$2^2 = 1$' and '$2^4 = 1$'. That the latter are invalid* is another matter. In the conditional context in which they occur, neither of them is asserted—even if the whole propositional complex should be expressed with assertive force. Thus we see that to the pseudo-propositions that are parts of a general theorem— of a theory—there correspond real propositions which occur in a proposition derived from the theorem by an inference from the general to the particular. If one of the propositions that occurs as an antecedent proposition is already recognized as valid, it may be omitted. Thus in the proposition 'If $1^2 = 1$, then $1^4 = 1$', the antecedent proposition may be left out once it is recognized as valid, so that we are left only with the consequent proposition '$1^4 = 1$'. In words, the transition may be expressed like this: Since $1^2 = 1$, therefore $1^4 = 1$. By means of such inferences we can get from a general theorem—a theory —to a particular proposition containing fewer antecedent-propositions. This procedure no doubt is called /382/ the application of a general theorem—a theory—to a particular case.

Perhaps, then, what Mr. Korselt calls a formal theory or purely formal system is a general theorem or theory in the sense in which I have just used these words. It would seem that Mr. Korselt has at least something like this in mind. I venture to say that much of his exposition will seem clearer if, when reading it, we bear in mind what was just said. He writes: "But modern mathematics, which more and more blends into exact logic,

---

* See p. 71 above for Frege's use of 'valid' and its derivatives in contexts like the present one. [*Trans.*]

no longer designates by its axioms (basic assertions) certain empirical facts—but at best *indicates* them; just as in algebra, a letter does not determine a number but merely indicates it."

It seems that Mr. Korselt has borrowed this usage of the word "indicate" from me. The comparison of axioms with letters is unfortunate, since in pure mathematics it is all one whether I use the letters '*a*', '*b*', '*c*' or the letters '*r*', '*s*', '*t*': each of the letters is considered to be simple. On the other hand, each of Hilbert's axioms is apparently supposed to have its own peculiarity which is based on its particular construction out of simple signs that can also occur in other combinations. On the other hand, the comparison of one of Hilbert's axioms with a pseudo-proposition such as '$a^2 = 1$' does not seem to be inappropriate. What Mr. Korselt probably has in mind I should express in the following way:

Contemporary mathematics—or let us just say, Mr. Hilbert—understands by an axiom not a real proposition which expresses a thought, but a pseudo-proposition from which, by an inference from the general to the particular, several real propositions can emerge which then do express thoughts.

My "expressing of thoughts" here corresponds to Korselt's "designating certain empirical facts." Thus Hilbert's axioms are parts of a general theorem that has a sense, although the parts themselves do not. And it is only as parts of a whole having a sense that those axioms are justified. They appear as antecedent propositions, or as we may also say, as presuppositions. And with this the following assertion by Mr. Korselt agrees very nicely:

> Nor is it the case that 'the ontological argument for the existence of God is brilliantly vindicated' by the axiom 'On every straight line there are at least two points'. For the former is intended to *prove* existence; the axiom *presupposes* it for all or at least some of the propositions that follow from it. In fact, the 'existence-propositions' of exact

logic and mathematics are no more than /383/ presuppositions of certain conditional propositions in whose 'assertion' certain concepts mentioned in the existence-
propositions no longer occur.

While not agreeing with every word, I nevertheless can detect
in this some corroboration of what I have just conjectured. An
axiom presupposes existence for all or some of the propositions
that follow from it. Very well! Then it is inseparable from them.
Neither the axiom nor the propositions that follow have a sense
of their own; rather, the axiom is an antecedent pseudo-proposition and these propositions that follow are consequent pseudo-
propositions; and these pseudo-propositions [together] form one
or several real propositions whose parts they are. For even Mr.
Korselt himself talks about antecedent propositions—these are
the real propositions—and their presuppositions—these are the
antecedent and consequent pseudo-propositions.

Of course, no refutation of my earlier view can be found in
this. When Mr. Korselt deviates in his usage of the words
"axiom" and "definition" from mine and attempts to construct
a refutation on the basis of this, then he refutes something I
have not said. One can give the appearance of refuting any
proposition whatever if one takes the liberty of understanding
the words in such a way that the proposition loses its import.

My train of thought was the following: If in the definition
of a concept of the first level it is permitted to mention existence
as a characteristic, then this may also be done in the definition
of the concept *God* which is of the first level; in which case the
existence of God would immediately follow. Now according to
Mr. Hilbert's understanding of the matter, the axiom in question
is part of the definition of a point, and existence is mentioned
in it as a characteristic. Consequently *that* right is claimed, which
in another case would permit the ontological argument. By pointing this out, I intended to induce Mr. Hilbert to reflect about
what he calls a definition. I assumed that he would recognize

that his use of the word is completely different from the customary one, and that he would perhaps go the way on which Mr. Korselt seems to have embarked, albeit without a clear understanding. Of course, this clarification and development for which I wanted to provide the impetus apparently did not come about at all in the case of Mr. Hilbert, and only incompletely in the case of Mr. Korselt. A definition in the traditional sense does not presuppose anything, but rather stipulates something. What I have said holds true if we understand the word "definition" as it has traditionally been understood in mathematics, and even if an axiom is part of a definition, /384/ as Mr. Hilbert wants it.[2] Now it is possible that Mr. Hilbert's procedure is nevertheless justified for a different sense of the word "definition"; but which sense are we to assume here? Let us try to get clear about this with the help of Mr. Korselt's pronouncements! To begin with, what Mr. Hilbert calls the definition of a point is certainly not a definition in the old sense of the word. Furthermore, the definition consists of axioms. These presuppose something and therefore undoubtedly are antecedent propositions, and pseudo-propositions at that. This close connection into which the words "axiom" and "definition" have thus been forced is quite foreign to their original usage. Therefore a definition, when understood in this sense, is no more than a whole consisting of several axioms connected by "and," where the latter themselves are pseudo-propositions (antecedent propositions). But then, no longer is there any real difference between a definition and an axiom. In which case a definition, too, would be an antecedent pseudo-proposition consisting of several pseudo-propositions connected by "and". It is immaterial whether several of the antecedent propositions are first combined into a whole which is then taken as an antecedent proposition, or whether the antecedent propositions are left uncombined. We can

2. Of course a principle in the old sense of the term neither presupposes nor stipulates something; rather, it asserts something. I use the term "principle" for any proposition that expresses an axiom.

see that in such a case, the so-called definitions are superfluous.

But let us investigate whether our conjecture about the nature of purely formal systems is further confirmed! Mr. Korselt writes, " 'Arithmeticized', or better, 'rationalized' mathematics merely arranges its principles in such a way that certain known interpretations are not excluded."

Here the principles will again be the antecedent pseudo-propositions of the general theorem. The word "interpretation" is objectionable, for when properly expressed, a thought leaves no room for different interpretations. We have seen that ambiguity simply has to be rejected and how it may appear to be necessary because of insufficient logical insight. I merely recall what I have said about the use of letters above, on p. 377. On the basis of our understanding of the nature of Korselt's purely formal system it is easy to guess what Mr. Korselt means by "interpretation." When we proceed from the general theorem "If $a > 1$, then $a^2 > 1$" to the particular one "If $2 > 1$, then $2^2 > 1$" by means of an inference, then the pseudo-proposition "$a > 1$" corresponds to the proper proposition '$2 > 1$'. According to Mr. Korselt's usage, '$2 > 1$' or the /385/ thought of this proposition will be an interpretation of '$a > 1$'. As if the general proposition were a wax nose which we could turn now this way, now that. In reality, we have not an interpretation but an inference.[3]

3. We may take this opportunity to illuminate Mr. Korselt's following pronouncement: "Propositions having the same wording should, if possible, be proved only once, even if they appear in different disciplines."

As if it were permissible to have different propositions with the same wording! This contradicts the rule of unambiguousness, the most important rule that logic must impose on written or spoken language. If propositions having the same wording differ, they can do so only in their thought-content. Just how could there be a single proof of different thoughts? This looks as though what is proved is the wording alone, without the thought-content; and as though afterwards different thoughts were then supposed to be correlated with this wording in the different disciplines. Rubbish! A mere wording without a thought-content can

In a pseudo-proposition there must occur signs that do not designate anything but merely indicate. Which signs are these in the present case? Clearly the words "point," "straight line," "lies in," "lies on," "lies between," etc. Therefore if, as Mr. Korselt would have it, Hilbert's geometry is a purely formal system, and if we have grasped the meaning of this expression correctly, then in Hilbert's geometry these words do not designate anything at all. And in fact, Mr. Korselt actually does say that the signs of formal theories have no reference whatever. Therefore the words "point," "plane," etc. are supposed to serve the purpose of lending generality of content to the theorem, as do the letters in algebra. And this once more agrees very well with what we determined above, namely that Hilbert's so-called definitions do not give references to these words. We also see it confirmed that these so-called definitions are not definitions, any more than the pseudo-proposition "$a > 1$" in the proposition "If $a > 1$, then $a^2 > 1$" is a definition. Letters intended to lend generality of content to a proposition receive no explanation, for they are not supposed to refer but merely indicate. Since letters are not supposed to be given references, definitions, whose purpose would be to given them, are here out of place. Sometimes what looks like an explanation of letters is really an antecedent proposition. And likewise in the present case. The words "point," "plane," etc. are used here like letters. /386/ What looks like an explanation of these words is an antecedent pseudo-proposition. Considered as a definition, it does not meet even the most modest requirements which a definition must meet. Since the term "antecedent proposition" is quite sufficient, I cannot see why the misleading words "definition" and "axiom,"

---

never be proved. What Mr. Korselt has in mind is, of course, the case where what is to be proved is a general proposition from which the propositions belonging to the different disciplines are derived by means of an inference from the general to the particular (or the less general). Here, too, we are prepared to hear Mr. Korselt talk of interpretation.

which traditionally have a different usage, should be used instead. What Mr. Hilbert calls a definition will in most cases be an antecedent pseudo-proposition, a dependent part of a general theorem.

Given this conception, it is not only pointless and inappropriate to demand of a formal theory that it give a determinate reference to its figures, which are formed on the model of proper names and concept-words; it is nonsensical to demand any reference of them at all. For they are not supposed to be designating, but merely indicating signs.

I do demand the solvability of a system of principles as to the unknowns occurring in it, and an unambiguous solution, if this system of so-called principles is supposed to be a definition that assigns references to the unknown signs. For this purpose can be achieved only when this demand is met. But in no way do I require that one should define everything. I certainly do not require that in order to define something, one construct a system of so-called principles. Least of all do I require that one explain signs which, like letters, are used only in an indicating and not a referring capacity; for that would be to demand nonsense.

Now when proceeding from a general theorem to a particular one by means of an inference, to every pseudo-proposition which is part of the former there corresponds a real one in the latter. These real propositions may indeed be principles—expressions of axioms—in the old and proper sense of that word. Since the axioms of Euclidean geometry are true, we may omit them wherever they occur as conditions. We then have made an application of the general theorem and have thus arrived at a proposition of Euclidean geometry. But other applications are also possible; Mr. Korselt mistakenly calls them interpretations.

We now understand how Mr. Hilbert's peculiar confusion in the use of the word "axiom" came about. The expression was transferred from the real to the pseudo-propositions which correspond to them. Through this misuse and that /387/ of the word "definition," insight into the logical nature of Hilbert's geometry has been made inordinately more difficult.

Let us continue in our investigation of Korselt's pronouncements! There we read, "In this way, *one* sequence of formal inferences can sometimes be 'interpreted' in *different* ways."

What can be interpreted is perhaps a sign or a group of signs, although the univocity of the signs—which we must retain at all cost—excludes different interpretations. But an inference does not consist of signs. We can only say that in the transition from one group of signs to a new group of signs, it may look now and then as though we are presented with an inference. An inference simply does not belong to the realm of signs; rather, it is the pronouncement of a judgment made in accordance with logical laws on the basis of previously passed judgments. Each of the premises is a determinate thought recognized as true; and in the conclusion, too, a determinate thought is recognized as true. There is here no room for different interpretations.

What is a formal inference? We may say that in a certain sense, every inference is formal in that it proceeds according to a general law of inference; in another sense, every inference is nonformal in that the premises as well as the conclusions have their thought-contents which occur in this particular manner of connection only in that inference. But perhaps the word "formal" is here supposed to be understood differently. Perhaps a series of formal inferences is not supposed to be a proper inference-chain but only the schema of one. Its interpretation would then consist in indicating an inference-chain that would proceed according to this schema. Now of what use is such a schema? Perhaps this, that in a given case we do not have to go through the whole inference-chain, but instead can pass directly from the first premises to the last consequent proposition. But then we no longer have a mere schema, but a general theorem.

Consider, for example, the following schema:

A is a *b;* every *b* is a *c;* therefore *a* is a *c;* therefore there is a *c!*

Here we obviously do not have an inference, for we do not have real propositions—no thoughts. But a chain of inferences can proceed according to this schema, and in accordance with the double occurrence of "therefore" it would consist of two inferences. The schema itself says nothing, but it provides the occasion for constructing propositions that do say something. To begin with the following:

/388/ If *a* is a *b,* and if every *b* is a *c,* then *a* is a *c;*
 If *a* is a *c,* then there is a *c.*

By means of an inference, we obtain from these the general theorem,

If *a* is a *b,* and if every *b* is a *c,* then there is a *c.*

In a given case, instead of proceeding in accordance with the schema of the inference-chain, we can use the general theorem in such a way that by means of an inference from the general to the particular we deduce from it a proposition which we can free from the conditions that have now been fulfilled. That, after all, is generally the use of a theorem: It keeps the result of a series of inferences ready for use whenever we wish. In this way we have again been led to something which Mr. Korselt probably calls a formal theory or a purely formal system.

But let us leave these abstract considerations and see how all this manifests itself in Hilbert's theory itself. If, as we have assumed, the words "point," "straight line," etc. do not designate but merely are to lend generality, like the letters in arithmetic, then it will be conducive to our insight into the true state of affairs to actually use letters for this purpose. Let us therefore stipulate the following: Instead of "the point *A* lies in the plane *a*," let us say, "*A* stands in the *p*-relation to *a*." Instead of "the point *A* lies on the straight line *a*," let us say "*A* stands in the q-relation to a." Instead of "A is a point," let us say, "A is a π."

Hilbert's Axiom I.1 can now be expressed like this:

> If A is a π and if B is a π, then there is something to which both A and B stand in the q-relation.

We must distinguish here between two generalities. The one engendered by the letters 'A' and 'B' is confined to this pseudo-axiom;[4] while the one engendered by 'π' and 'q' extends to a general theorem (purely formal system, formal theory) of which this pseudo-axiom is only a dependent part that is without meaning on its own.

/389/ Hilbert's Axiom I.6 (I.5)[5] can be expressed like this:

> If A is different from B, if A and B stand in the p-relation to a, and if A, B, and C stand in the q-relation to a, then C stands in the p-relation to a.

The generality engendered by the letters 'A', 'B', 'C', 'a', and 'a' is limited to this pseudo-axiom.

Hilbert's Axiom I.7 (I.6)[5] now looks like this.

> If A stands in the p-relation to both a and β, then there exists something distinct from A which stands in the p-relation to both a and β.

The generality engendered by the letters 'A', 'a', and 'β' is limited to this pseudo-axiom.

We still need a pseudo-axiom Σ, which Mr. Hilbert does not have and which we express like this:

4. This expression may occasion objections. One could say that generality, after all, belongs to the thought-content of a proposition; hence how can there be talk of it here, in the case of a pseudo-proposition that does not even express a thought? It is to be understood in this way: the generality engendered by 'A' and 'B' is supposed to apply to the content of every real proposition arising out of this pseudo-axiom by replacing the letters 'π' and 'q', which merely indicate, by determinate signs.

5. What is in parentheses refers to the first edition.

Σ. If *A* stands in the *p*-relation to α, then *A* is a Π.

Here the generality engendered by the letters '*A*' and 'α' is limited to this pseudo-axiom. The letters 'Π', '*p*', and '*q*' neither refer to something, nor are they supposed to lend generality to the content of the particular pseudo-axioms; wherefore the latter, taken in themselves, do not express thoughts but are without sense. It is because of this that I add the "pseudo." For in the case of proper principles we must have thoughts. To be sure, the letters 'Π', '*p*', and '*q*' are supposed to engender generality, but the latter is supposed to extend over a theorem whose antecedent pseudo-propositions are these pseudo-axioms.

It seems to me that one advantage in recasting Hilbert's pseudo-axioms in this way is immediately apparent; namely, that no one will imagine that he understands such a pseudo-axiom, or that he finds a thought expressed in it. In reality, however, nothing essential has been changed by the use of letters in place of the expressions "point," "lies in," "lies on." At least, not as long as these expressions do not refer to anything, but like the letters, merely lend generality to the purely formal system, to use Mr. Korselt's phrase.[6] Therefore if these pseudo-axioms, when recast in the Hilbertian form, give the appearance of having a sense, then clearly this is because being familiar with Euclidean geometry, we are /390/ used to associating a sense with the words "point," "lies in," etc., and because we do not forget the latter as we should when concerning ourselves with Hilbert's "Foundations." As a matter of fact, we must here assume the outlook of someone who has never heard anything of points, planes, etc.; and in this we are not very successful. We have much better success with signs with which we in fact

6. But even if Mr. Hilbert's axioms were supposed to give references to the words "point," etc., nothing essential would be changed, for these very references would now be given to the letters 'π', etc. If these letters do not obtain a reference by this move, then neither do the words.

have not as yet associated a sense. The two cases are, however, essentially the same.

From the fact that the pseudo-axioms do not express thoughts, it further follows that they cannot be premises of an inference-chain. Of course, one really cannot call propositions—groups of audible or visible signs—premises anyway, but only the thoughts expressed by them. Now in the case of the pseudo-axioms, there are no thoughts at all, and consequently no premises. Therefore when it appears that Mr. Hilbert nevertheless does use his axioms as the premises of inferences and apparently bases proofs on them, these can be inferences and proofs in appearance only.

Now we could try the following move. The inferences may be conducted purely formally, as if the letters 'Π', 'p', and 'q' did refer; for after all, what they might refer to is all one so far as the correctness of the inference is concerned. If for these letters we now substitute actually referring signs of such a kind that true propositions thereby emerge from the pseudo-axioms, then a true proposition will emerge from the so-called consequent proposition as well.

Let us try this with an example! First, from our pseudo-axioms I.1 and Σ we derive the pseudo-proposition

> If A as well as B stands in the p-relation to a, then there is something to which A as well as B stands in the q-relation.   (A)

Furthermore, by combining I.6 with itself, we derive the pseudo-proposition

> If A is different from B, and if A as well as B stands in the p-relation to a as well as to β; and if A, B, and C stand in the q-relation to a, then C stands in the p-relation to a as well as to β.   (B)

From this we further derive the pseudo-proposition

If A is different from B, and A as well as B stands in the p-relation to a as well as to β, and if there exists something to which A as well as B stands in the q-relation, then there is an object such that whatever stands in the q-relation to it stands in the p-relation to a as well as to β. (Γ)

/391/ From this and (A), which was just derived above, we then obtain the pseudo-proposition

If A is different from B, and A as well as B stands in the p-relation to a as well as to β, then there is an object such that whatever stands in the q-relation to it stands in the p-relation to a as well as to β. (Δ)

From this we further infer the pseudo-proposition

If A stands in the p-relation to a as well as *to* β, and if there is something distinct from A such that it stands in the p-relation to a as well as to β, then there is an object such that whatever stands in the q-relation to it stands in the p-relation to a as well as to β. (E)

When we combine our pseudo-axiom I.7 with this, we obtain

If A stands in the p-relation to a as well as to β, then there is an object such that whatever stands in the q-relation to it stands in the p-relation to a as well as to β. (Z)

And from this we further obtain the pseudo-proposition

If there is something that stands in the p-relation to both a and β, then there is an object such that whatever stands in the q-relation to it stands in the p-relation to a as well as to β. (H)

In Mr. Hilbert's writings, we find the following wording for the above:

Two planes have either no point or a staight line in common.

Without becoming exercised over the fact that our wording is significantly longer than Hilbert's, let us ask whether what we have just constructed is really an inference-chain. Clearly not; for the links are only pseudo-propositions, as is the consequent proposition. Not one of them contains a thought. But if instead of our letters we were to use Hilbert's words "point," "lies in," etc., the propositions we should thus obtain would not have a sense either—assuming, of course, that these words are to have no more sense than do our letters. But then, what is the point of all these merely apparent inferences? What is the point of going through all these pseudo-propositions if the proposition that we finally arrive at is as much without sense as the preceding ones? Well, let us recall that although pseudo-propositions by themselves do not express thoughts, nevertheless they may be constituents of a whole that does have a sense. We cannot treat our pseudo-axioms as independent propositions that contain true thoughts and hence can serve as the foundations of our logical constructions; rather, we must carry them along as antecedent /392/ pseudo-propositions. Therefore, instead of our pseudo-proposition (A), we now have to write:

If it holds universally of A and a that

if A stands in the p-relation to a, then A is a II;

and if it holds universally of A and B that

if A is a II and if B is a II, then there is something to which both A and B stand in the q-relation;

then it holds universally of A, B, and a that

A as well as B stands in the p-relation to a, then there is something to which A as well as B stands in the q-relation.

The preceding is a proposition that does express a thought; nor do we find more than one thought in it. Those parts of it that present themselves grammatically as propositions are merely pseudo-propositions. The letters 'Π', '*p*', '*q*' lend generality of content to the whole proposition, while the generality effected by the letters '*A*', '*B*', '*a*' is always restricted to one of the indented component pseudo-propositions.[7] From this it becomes apparent how the component pseudo-propositions, although senseless in and by themselves, nevertheless can form a proposition that expresses a thought.

Similarly, in order to obtain real propositions we shall have to supplement the remaining pseudo-propositions that occur in the apparent inference-chain with our pseudo-axioms as antecedent propositions. From our merely apparent inference-chain, we obtain a real one. Only then do we have real premises and real consequent propositions. Thus in the end our consequent pseudo-proposition (H) will also have to be supplemented by our four pseudo-axioms, which will appear here as antecedent pseudo-propositions. And thus we shall obtain a consequent proposition that really does contain a thought. To be sure, this consequent proposition will be rather lengthy, but it is only through such supplementation that we obtain complete insight into the logical connection. Therefore let us not be distressed by the amount of effort involved in constructing such a proposition, for it appears that many confusions in mathematics are caused by an unnecessary economy with printer's ink and by false elegance. The principle of achieving as much as possible with the least possible means, if correctly understood, is certainly to be applauded; only the means must not be gauged by the consumption of printer's ink. In place of our previous pseudo-proposition (H), we now obtain the following consequent proposition: /393/

If it holds universally of *A* and *a* that

if *A* stands in the *p*-relation to *a*, then *A* is a Π,

and if it holds universally of *A* and *B* that

7. Cf. note 4 above.

if A is a II and if B is a II, then there is something to which both A and B stand in the q-relation;

also, if it holds universally of A, B, C, a, and a that

if A is distinct from B, if A and B stand in the p-relation to a, and if A, B, and C stand in the q-relation to a, then C stands in the p-relation to a;

furthermore, if it holds universally of A, a and β that

if A stands in the p-relation to a as well as β, then there is something distinct from A that stands in the p-relation to both a and β;

then it holds universally of a and β that

if there is something that stands in the p-relation to both a and β, then there is an object such that whatever stands in the q-relation to it stands in the p-relation to both a and β.

The indented pseudo-propositions are in part our pseudo-axioms and in part our previous consequent pseudo-proposition, and we can see they are merely dependent parts of the real consequent proposition which alone expresses a thought. The generality effected by the letters 'A', 'B', 'C', 'a', 'a', and 'β' is in each case restricted to the indented component pseudo-proposition in which they occur; while the generality effected by 'II', 'p' and 'q' extends to the whole proposition. We now see what really has been proved. This is not at all evident from Mr. Hilbert's consequent proposition:

Two planes have either no point or a straight line in common.

For here, unless we are to understand the words in the Euclidean sense, just about everything is unknown.

Let us continue to examine Mr. Korselt's expositions as to

whether or not they agree with our conception of his formal theory. Thus, on p. 403 we read, "We must therefore distinguish between those formal theories ('purely formal systems') that can be related to other experiences and those for which as yet no such correlation is known."

What Mr. Korselt here calls "relating to other experiences" and "correlation" is evidently the same as what he previously had called "interpreting" and "interpretation," and is nothing but an inference /394/ from the general to the particular. Therefore, if by means of such an inference it is possible to go from a general theorem to a particular one of such a kind that the antecedent propositions in it which correspond to the antecedent pseudo-propositions of the general proposition are real and indeed valid propositions, then Mr. Korselt will say that the formal theory (our general theorem) can be related to other experiences; that it can be correlated to other experiences. This would also seem to be the objectiveness of which he speaks in the following: "The 'objectiveness' and above all the consistency of a purely formal system is always and necessarily demonstrated by exhibiting objects of which the basic assertions hold."

We can now understand this somewhat. For example, let the following be a general theorem (a formal theory, a formal system):

If a is a square root of 1, then a is a fourth root of 1.

By means of an inference from the general to the particular, we derive from it the following:

If 1 is a square root of 1, then 1 is a fourth root of 1.

And according to Mr. Korselt's manner of speaking, we are herewith presented with an object, to wit 1, of which the basic assertion holds. For after all, 1 is a square root of 1. And in Mr. Korselt's words, with this the objectiveness and, even more, the consistency of our purely formal system has been proved.

In the previously considered example taken from Hilbert's geometry, matters are of course not quite so simple, since not just one but several antecedent pseudo-propositions occur, and because not just one but three letters ('Π', 'p', 'q') appear. Furthermore we have this difference, that these letters do not indicate objects—or, as we could also say, substitute for proper names—but partly concepts, partly relations. Consequently, since these so-called basic assertions cannot fit objects, it cannot be objects that are presented here but only a concept and relations. However, an inference from the general to the particular is possible in this way: that 'point' stands in the place of 'Π', 'lies in the plane' stands in the place of 'stands in the p-relation to', 'lies on the straight line' stands in the place of 'stands in the q-relation to', where these expressions are to be understood in the Euclidean sense. From our pseudo-axioms we thus derive real principles which do express axioms; /395/ and this is the presentation of a concept and relations by means of which, in Mr. Korselt's words, the objectiveness[8] of our purely formal system (general theorem) is proved. At this point however, it must be emphasized that our theorem has been proved and is true quite independently of the proof of this so-called objectiveness. What Mr. Korselt says is therefore quite correct: we should not from the very start demand objectiveness of a purely formal system. The following statement also somehow hits the mark, even though in it there seems to have occurred a confusion to which we shall have to return later: "Even if we then call the latter (the purely formal system) 'an empty playing with words, signifying nothing' and the like, as a strictly lawlike connection of propositions, it has no further need of any special 'dignity.' "

Indeed, a general theorem needs no greater dignity than that accruing to it because it expresses a true thought. The only thing

8. Of course, this word is not quite appropriate in the present context.

to which I take exception here is the expression "of propositions". By a real proposition I understand the expression of a thought, and therefore something sensible; a sequence of words that can be heard, or a group of signs that can be seen. This last holds true even of pseudo-propositions. A concatenation of real and pseudo-propositions will belong to a grammar. Mr. Korselt does not distinguish strictly between the external or sensible, and the thought-content. In the present case, he probably means a concatenation of thoughts. But even this cannot be what is presently at issue; for here we have pseudo-propositions, none of which have a sense. Therefore what is meant here needs still more accurate expression, where it would of course be necessary to coin a new phrase.

As for the rest, I agree with it, particularly with the claim that objectiveness in the sense previously indicated is not necessary to ensure a thought-content for the proposition. Matters would of course lie differently if the words "point," "lying in," etc. were not to be used indicatively to lend generality to the content of a proposition, as are letters, but rather [were to be used] as referring concept- and relation-words. In that case, of course, there would have to be a concept that would be designated by the word "point" and a relation that would be designated by the expression "lies in."

But it seems to me that here a confusion obtains. The charge of being an empty playing with signs may justifiably be leveled against certain formal theories; these, however, are quite different from general /396/ theorems of the kind considered here. For in the latter we always have a sense. Those other formal theories, however, proceed after the manner of Dr. Ironbeard.* Since the sense occasions difficulties now and then, it is simply

---

* Dr. Ironbeard—*Eisenbart* in German—is the traditional German figure of a quack who effects cures no matter what, even at the price of killing the patient. Hence the expression, "nach der Methode des Dr. Eisenbarts verfahren." [*Trans.*]

exorcised. What remains is of course the inanimate sign. The originator of such a theory does not want to express thoughts with his signs, but merely wants to play with them according to certain rules. Consequently, it cannot be truth that is here at issue. The word "theory" is really quite inappropriate; we ought to say "game." At least, it would be so if the execution of the enterprise were consistent. But this is never the case: the theorists want to have their cake and eat it too. They empty the signs so as to escape inconvenient questions; but then they refuse to acknowledge that the signs are really empty. Thus they become entangled in a thicket of contradictions. What we have hitherto called a formal theory is something quite different. To be sure, we also use signs that have no references, but they contribute to the expression of thoughts in the familiar way. To express a thought by using letters alone and not using any referring signs is impossible. In those unacceptable so-called formal theories, signs like '$1/2$', '$\sqrt[3]{5}$' neither are referring proper names, as is usually the case, nor serve to lend generality of content to a proposition, as do the letters. They are not means of understanding and communication; rather, they are objects with which one plays according to certain rules. Mr. Korselt confounds these two quite distinct cases because he has not clearly understood the peculiarity of either. He finds a thought expressed in an arrangement of chess pieces. Perhaps he has some mental equipment for grasping thoughts that I lack. Does a whole new realm of thoughts here open up before us? Very interesting! But I cannot quell certain misgivings. If Mr. Korselt means that an arrangement of chess pieces expresses a thought merely because of the rules of chess, then I question this and shall continue to question it until I am presented with this thought as expressed in English; until I am shown how, in virtue of the rules of chess, this thought is expressed by that arrangement. After all, laying down rules for the use of chess pieces and stipulating the reference of a sign are at first glance two quite distinct things. If someone really thinks that under certain conditions the former could be combined with the latter, then he has to prove this. To my knowledge, however, no one has as yet even tried to do so.

/397/ If an axiom contains a hitherto unknown sign, then according to Mr. Korselt it is a rule for the use of this sign. If we cannot agree on the reference of a sign, then we should acquire one or several more propositions about the sign or using the sign. From this Mr. Korselt concludes that " 'The sign has no reference' will therefore mean 'We are not acquainted with any propositions regulating the use of this sign in general or for a given domain'." How he arrives at this conclusion escapes me, since nowhere in the preceding has there been any talk of regulating the use of a sign. But this much is clear: that in some admittedly mysterious way, such regulating is supposed to give a reference to the sign. Let us take an example! The axiom "Every anej bazet at least two ellah" regulates the use of the words "anej," "bazet," and "ellah." If in spite of this we nevertheless did not agree on the reference of the word "anej," this would be an indication that one of us should acquire more propositions about or using the word "anej." In such a case I should be more than happy to provide more propositions using the word "anej"; and if necessary, even propositions about this word.

The rule "Every anej bazet at least two ellah" deserves to be followed most conscientiously by all modern mathematicians. But I perceive a voice from the lower depths: "How can that be a rule! After all, in a rule something must be demanded, ruled out, or permitted. I expect imperatives, or words like 'must', 'ought', 'may', 'is ruled out', etc. Nothing of the kind do I find in this so-called rule. It certainly seems as if what is being talked about is a concept designated by the word 'anej'. But if a rule about the use of the word 'anej' is supposed to be found in this peculiar, proposition-like construction, then it is the word itself which is the subject of the discussion. Principles and theorems are propositions using a sign; rules are propositions about a sign."

To this I should like to reply, "How do you know, kind sir, that the word 'rule' is being used here in the sense familiar to you? And even if it is, it is quite understandable that in a sausage mixture such as the one we have before us, the peculiarity of a particular constituent is no longer clearly recognizable."

At times Mr. Korselt takes unnecessary pains, e.g. in the case of the domain of application of the axioms. If an axiom is a rule about the use of the unknown signs occurring in it, then /398/ the latter naturally form the domain of application of the rule and therefore of the axiom. Thus the question has been answered in the simplest terms.

If we use "axiom" in the Euclidean sense, then contrary to what Mr. Korselt assumes, there cannot be invalid axioms. In the case before us, however, an invalid axiom containing unknown signs will be an invalid rule for the use of these signs. It would be impertinent to ask what purpose such a rule might serve.

Herewith we shall leave these so-called rules and what is connected with them. If we strike it from Korselt's expositions, we free them of confusions.

As far as consistency and independence are concerned, it is Mr. Hilbert's opinion that they are to be proved of the axioms. Here there arises the question whether he means his pseudo-axioms or the axioms in the old Euclidean sense.

Mr. Korselt writes:

It is irrelevant whether it is the axioms or the characteristics of the concepts introduced that are said to be consistent. The former corresponds more closely to ordinary usage, according to which two propositions are called 'independent' of one another if under certain circumstances both, under other circumstances not both, obtain; whereas they are called 'incompatible' if there are no conditions under which both are satisfied together.

What does "the proposition obtains" mean? Surely that the proposition expresses a true thought. Now a real proposition expresses a thought. The latter is either true or false: *tertium non datur*.[9] Therefore that a real proposition should obtain under certain circumstances and not under others could only be the case if a proposition could express one thought under certain circumstances and a different one under other circumstances. This, however, would contravene the demand that signs be unambiguous—[a demand] to which we must adhere under all circumstances, as has been argued at length above. A pseudo-proposition does not express a thought at all; consequently we cannot say that it obtains. Therefore it simply cannot happen that a proposition obtains under certain circumstances but not under others, whether it be a real proposition or a pseudo-proposition.[10] Still, we can guess /399/ what Mr. Korselt means. It can concern only pseudo-propositions, and here comes Mr. Korselt once again with his interpretations. He interprets a proposition like this, and it obtains; he interprets it otherwise, and it does not obtain. He turns the wax nose now to the right, now to the left, just as he pleases. For example, let us take the proposition "On a straight line there are at least two points!" Now let us interpret the word "point" as foot, the words "straight

9. For we are here in the realm of science. In myth and fiction, of course, there may occur thoughts that are neither true nor false but just that: fiction.
10. Even when Mr. Korselt says, "But on the other hand, a formal theory may be applied to a given domain only if we have assured ourselves of the correctness of the principles for that domain," he seems to mean that one and the same proposition could hold in one domain but not in another. What Mr. Korselt means is expressed more precisely like this: "If we have derived a particular proposition from a general one by an inference from the general to the particular, then we may omit the antecedent propositions in it if and only if these are valid."

line" as worm, and the words "there are" as has. We then interpret our proposition thus: A worm has at least two feet. Almost as easily as we have here obtained something false, can we obtain something true from this proposition by means of different interpretations. We now see how right Mr. Korselt is when he says that a proposition may hold under some circumstances, but not under others; it simply all depends upon the interpretation. But let us stop joking. A proposition that holds only under certain circumstances is not a real proposition. However, we can express the circumstances under which it holds in antecedent propositions and add them as such to the proposition. So supplemented, the proposition will no longer hold only under certain circumstances but will hold quite generally. The original proposition appears in it as a consequent proposition; and as a pseudo-proposition at that. Well, however we may consider the matter, the upshot is one and the same: If we suppose that a proposition can hold under certain circumstances but not under others, then we allow ourselves to be led by the nose by self-induced inexactitudes of expression.

In the passage cited, Mr. Korselt says, "It is irrelevant whether it is the axioms or the characteristics of the concepts introduced that are said to be consistent."

To be sure, irrelevant for one who cares nothing for precision of expression and who is not concerned to get a deeper insight into the state of affairs. Axioms are simply not characteristics of concepts. Therefore from the very first the consistency of the axioms must be distinguished from the consistency of the concepts introduced. Whoever maintains that there is no difference must prove it. The mere assertion by Mr. Korselt does not suffice to establish this lack of difference as an assured tenet of science. '4^2' and '2^4' must first be distinguished; only after the coincidence of their references has been proved can they be interchanged.

/400/ Mr. Korselt continues, "The former corresponds more closely to ordinary usage, according to which two propositions

are called 'independent' of one another if under certain circumstances both, under other circumstances not both, obtain."
    Really? Is this ordinary usage? In the propositional complex
"If $a > 1$, then $a > 0$" we have two pseudo-propositions, '$a > 1$'
and '$a > 0$'. Are these called independent of one another in
ordinary usage? And yet according to Mr. Korselt's manner of
speaking—which must of course be rejected—we have circumstances under which both hold ($2 > 1$ and $2 > 0$, $3 > 1$ and
$3 > 0$), and other circumstances under which it is not the case
that both hold ($1 > 1$ and $1 > 0$, $0 > 1$ and $0 > 0$). This much
may perhaps be admitted, that according to ordinary usage
'$a > 1$' is called independent of '$a > 0$'. But ordinary usage cannot decide anything for one who does not want to let himself
be deceived by words but rather wants to get to the bottom of
the matter. For there is always this question: Is ordinary usage
appropriate to the occasion?
    Therefore we cannot agree with Korselt's explanation of the
independence of propositions for a variety of reasons. But this
much is clear—that the former is only supposed to concern
pseudo-propositions. Therefore according to the opinion of Mr.
Korselt, the question of independence will concern not axioms
in the Euclidean sense, but rather the pseudo-axioms of Mr.
Hilbert. And in this he is probably quite correct, since real
axioms very likely do not have a place in Mr. Hilbert's presentation at all.
    What, then, are we really talking about when, for example,
we call the propositions '$a > 0$' and '$a > 1$' independent of one
another? Is it the groups of signs, irrespective of whether these
signs have a sense? Of course not! On the other hand, neither
the group of signs '$a > 0$' nor '$a > 1$' has a sense. Still, '$a > 0$'
is not completely divorced from all sense, since it can belong
to a larger whole that does express a thought, and also because
it itself contains a referring part. The largest part of '$a > 0$' that
still refers is the predicative one, and its reference is the concept
of a positive number. Similarly, the reference of the largest part

of '$a > 1$' that still refers is the concept of a number smaller than
1. Now it may be conjectured that when it is apparently asserted
of these propositions that they are independent of one another,
something is really being said about these concepts. And indeed,
one generally supposes that the first concept is not subordinate
to the second, nor the second to the first. The point may also be
expressed like this: "Some positive numbers are not smaller than
1, and some numbers that are smaller than 1 are /401/ not
positive." From this we can clearly see that we are here concerned with relations between concepts.

A proposition in which, following Mr. Korselt's unquestionably imprecise usage, one proposition is presented as being
dependent upon others, consists of a consequent pseudo-proposition and one or more antecedent pseudo-propositions. They
are pseudo-propositions because in them occur constituents that
do not designate anything but only indicate so as to lend generality of content to the whole proposition. This proposition,
insofar as it holds true, will be what according to Mr. Korselt
should be called a purely formal system or formal theory. But
when this proposition does not hold and therefore independence
rather than dependence obtains, the thought of the whole proposition is to be denied. Therefore a proposition in which an
independence of this kind is asserted denies the validity of such
a general proposition. Therefore the antecedent pseudo-propositions are not brought into relation with the consequent pseudo-
propositions by a general proposition such as the preceding; nor
even the thoughts that they might express, for there are none.
Rather, what is brought into relation are the references of parts
of the pseudo-propositions; and these parts do not express
thoughts. What has just been said about general propositions
also holds of their negations, i.e. of propositions in which, according to Mr. Korselt's usage, what is supposed to be asserted
is the independence of one proposition from others. Therefore
even this sort of proposition relates neither propositions nor
thoughts, but instead relates the references of the parts of

pseudo-propositions. It is these pseudo-propositions that are the pseudo-axioms of Mr. Hilbert's independence-proofs. From this it follows that the independence proved by Mr. Hilbert does not concern these pseudo-axioms.

When one uses the phrase "prove a proposition" in mathematics, then by the word "proposition" we clearly mean not a sequence of words or a group of signs, but a thought; something of which one can say that it is true. And similarly, when one is talking about the independence of propositions or axioms, this, too, will be understood as being about the independence of thoughts. It is therefore not at all unnecessary to say that this conception is false and to reject the manner of speaking that gives rise to this misunderstanding.

We have to distinguish between the external, audible, or visible which is supposed to express a thought, and the thought itself. It seems to me that the usage prevalent in logic, according to which only the former is called a proposition, is preferable. Accordingly, we simply cannot say that one proposition is independent of other propositions; /402/ for after all, no one wants to predicate this independence of what is audible or visible. Since pseudo-propositions, and hence also pseudo-axioms, have no thought-content, something can be asserted of them only when they are understood in the last sense recommended above; but it is precisely independence that cannot be predicated of them.

Summarizing our result, we may say this: Hilbert's independence-proofs concern neither the independence of propositions in the sense just recommended, nor the independence of thoughts. Rather, they concern the independence of the references of the parts of pseudo-propositions. These parts are the largest that still refer. But they are not propositions, and therefore do not express thoughts. We lack a short designation for the references of such parts, a designation covering all cases. In the simpler cases, we have concepts (*positive number, number smaller than 1*). Toward the end of my first essay[11] I accommo-

11. *Jahresbericht, 12,* 323 and 324. [pp. 27 ff. above. Trans.]

dated my usage to one suggested in a letter by Mr. Hilbert and
called them characteristics, although this does not quite agree
with my own usage.

It must be noted that Mr. Hilbert's independence-proofs simply are not about real axioms, the axioms in the Euclidean
sense; for these, surely, are thoughts. Now nowhere in Mr. Hilbert's writings do we find a differentiation that might correspond
to our own between real and pseudo-propositions, between real
and pseudo-axioms. Instead, Mr. Hilbert appears to transfer the
independence putatively proved of his pseudo-axioms to the
axioms proper, and that without more ado, because he simply
fails to notice the difference between them. This would seem
to constitute a considerable fallacy. And all mathematicians who
think that Mr. Hilbert has proved the independence of the real
axioms from one another have surely fallen into the same error.
They do not see that in proving this independence, Mr. Hilbert
is simply not using the word "axiom" in the Euclidean sense.
The fault here lies in the double usage of the words "point,"
"straight line," etc., which on the one hand, like letters, are to
lend generality to the whole theory, in which case they do not
designate anything; and on the other hand have their traditional
references in the Euclidean axioms. In the former case his
axioms are merely pseudo-axioms without sense, since only the
whole (the formal theory, the purely formal system of Mr.
Korselt) whose /403/ dependent parts they are, has a sense—
in which case the Euclidean axiom of parallels simply does not
occur, and consequently nothing is proved of it. In the other
case real axioms do occur. But then these independence-proofs
are inappropriate, since it is impossible to substitute other concept-words for "point," "straight line," etc. But surely it is on
this very possibility that such a proof depends.

Even Mr. Korselt seems to have overlooked this comple-
mentary side of the matter when he emphasizes the difference
of the axioms of modern mathematics from the Euclidean ones.

This difference, after one has used its advantages in the beginning, cannot be denied in the final result.

The twilight which rules Hilbert's presentation must first give way to uniform illumination before the matter can become clear. And then the mixing of axioms and definitions will also come to an end.

III.

/423/ Now the question may still be raised whether, taking Hilbert's result as a starting point, we might not arrive at a proof of the independence of the real axioms.

We must first ask what is here to be understood by 'independence'; for what is called independence in the case of pseudo-propositions—even if this is an inexactitude—cannot enter into consideration here. If we take the words "point" and "straight line" in Hilbert's so-called Axiom II.1 in the proper Euclidean sense, and similarly the words "lie" and "between," then we obtain a proposition that has a sense, and we can acknowledge the thought expressed therein as a real axiom. Let us designate it by "[II.1]"! Let [II.2] emerge in a similar way from Hilbert's II.2. Now if one has acknowledged [II.1] as true, one has grasped the sense of the words "point," "straight line," "lie," "between"; and from this the truth of [II.2] immediately follows, so that one will be unable to avoid acknowledging the latter as well. Thus one could call [II.2] dependent upon [II.1]. Of course, we do not have an inference here; and it seems inexpedient to use the word "dependent" in this way, even though linguistically it might be possible [to do so].

What I understand by independence in the realm of thoughts may become clear from the following. I here use the word "thought" instead of "proposition," since surely it is only the thought-content of a proposition that is relevant, and the former is always present in the case of real propositions—and it is only with these that we are here concerned. What I call a group of thoughts will be apparent from the fact that I say: The linguistic expression of a group of thoughts consists of real propositions connected by "and." We can think of a group of thoughts as one thought constituted of other thoughts.

Let Ω be a group of true thoughts. Let a thought G follow from one or several of the thoughts of this group by means of a logical inference such that apart from the laws of logic, no proposition not belonging to Ω is used. Let us now form a new group of thoughts by adding the thought G to /424/ the group Ω. Call what we have just performed a logical step. Now if through a sequence of such steps, where every step takes the result of the preceding one as its basis, we can reach a group of thoughts that contains the thought A, then we call A dependent upon group Ω. If this is not possible, then we call A independent of Ω. The latter will always occur when A is false.

In § 10 of his "Foundations of Geometry," Mr. Hilbert raises the question whether axioms are independent of one another, and then continues: "Indeed, it turns out that none of the axioms can be deduced from the remaining ones by means of logical inferences." According to this, he appears to use the word "independent" just as has been stipulated above. But apparently it only seems that way, since in our case we are concerned with thoughts; Mr. Hilbert's axioms, however, are pseudo-propositions which therefore do not express thoughts. This may be seen from the fact that according to Mr. Hilbert an axiom now holds, and now does not. A real proposition, however, expresses a thought, and the latter is either true or false; *tertium non datur*.[12] A false axiom—where the word "axiom" is understood in the proper sense—is worthy of exhibition in Kastan's Waxworks,* alongside a square circle. Moreover, it appears that Mr. Hilbert's usage vacillates. If something is supposed to express now this

12. Cf. note 9 above.
* A waxworks similar to Mme. Tussaud's. [*Trans.*]

thought, now that, then in reality it expresses no thought at all. Hilbert's pseudo-axioms are of this nature. They are groups of sounds or written signs which are apparently intended to express thoughts without, however, actually doing so. Now it is clear that such groups cannot be premises of inferences, for inferring is not an activity within the realm of the sensible. Therefore—strictly speaking—not even real propositions can be premises of inferences, but at best the thoughts expressed by them. Deducing something by logical inferences from Hilbert's pseudo-axioms is about as possible as cultivating a garden plot by means of mental arithmetic. In this case, therefore, Mr. Hilbert must not be using the word "axiom" in the usual way. With such vacillations in usage, it cannot be ascertained how he understands the word "dependent." He himself probably knows this as little as he knows what he means by "axiom."

By way of elucidating my explanation, let me add a few remarks.

/425/ In taking a logical step from the thought-group Ω, we are applying a logical law. The latter is not to be counted among the premises and therefore need not occur in Ω. Thus there are certain thoughts, namely the laws of logic, which are not to be considered when dealing with questions concerning the dependence of a thought.

Only true thoughts can be premises of inferences. Therefore if a thought is dependent upon a thought-group Ω, then all the thoughts in Ω that are used in the proofs must be true. But, one might perhaps object, surely one can make deductions from certain thoughts purely hypothetically without adjudging the truth of the latter. Certainly, purely hypothetically! But then it is not these thoughts that are the premises of such inferences. Rather, the premises are certain hypothetical thoughts that contain the thoughts in question as antecedents. Even in the final result, the thoughts in question must occur as conditions; whence it follows that they were not used as premises, for otherwise they would have disappeared in the final result. If one has left them out, one has simply made a mistake. Only after one of the thoughts in question has been admitted as true can one omit it as an antecedent. This occurs through an inference having as one of its premises the thought now admitted as true.

We cannot from the very start exclude the case where every thought of group Ω' is dependent upon group Ω, while at the same time every thought of group Ω is dependent upon group Ω'. Therefore from the fact that all thoughts of group Ω' are dependent upon Ω, it cannot be concluded without more ado that the thoughts of group Ω are independent of Ω'.

Another case that cannot be ruled out a priori is that in which a thought A is dependent upon both a group Ω as well as upon a group Ω_1, whereby no thought of group Ω is dependent on Ω_1, nor any thought of group Ω_1 dependent upon Ω. In this case, therefore, A can be proved from group Ω_1 without the thoughts of group Ω even being known. In spite of this we could not call A independent of group Ω.

We now return to our question: Is it possible to prove the independence of a real axiom from a group of real axioms? This leads to the further question: How can one prove the independence of a thought from a group of thoughts? First of all, it may be noted that with this question we enter into a realm that is otherwise foreign to mathematics. For although like all other disciplines /426/ mathematics, too, is carried out in thoughts, still, thoughts are otherwise not the object of its investigations. Even the independence of a thought from a group of thoughts is quite distinct from the relations otherwise investigated in mathematics. Now we may assume that this new realm has its own specific, basic truths which are as essential to the proofs constructed in it as the axioms of geometry are to the proofs of geometry; and that we also need these basic truths especially to prove the independence of a thought from a group of thoughts.

To lay down such laws, let us recall that our definition reduced the dependence of thoughts to the following of a thought from other thoughts by means of an inference. This is to be understood in such a way that all these other thoughts are used as premises of the inference and that apart from the laws of logic no other thought is used. The basic truths of our new discipline which we need here will be expressed in sentences of the form:

> If such and such is the case, then the thought G does not follow by a logical inference from the thoughts A, B, C.

Instead of this, we may also employ the form:

> If the thought G follows from the thoughts A, B, C by a logical inference, then such and such is the case.

In fact, laws like the following may be laid down:

> If the thought G follows from the thoughts A, B, C by a logical inference, then G is true.

Further,

> If the thought G follows from the thoughts A, B, C by a logical inference, then each of the thoughts A, B, C is true.

For we have seen that only true thoughts can be the premises of inferences. But our aim is not to be achieved with these basic truths alone. We need yet another law which is not expressed quite so easily. Since a final settlement of the question is not possible here, I shall abstain from a precise formulation of this law and merely attempt to give an approximation of what I have in mind. One might call it an efflux of the formal nature of logical laws.

Imagine a vocabulary: not, however, one in which words of one language are opposed to corresponding ones of another, but where on both sides there stand words of the same language but having different senses. Let this occur in such a way that proper names are once again opposed to proper /427/ names and concept-words again to concept-words. Furthermore, let this occur with preservation of level,[13] so that to words for first-level concepts on the left there correspond similar ones on the right. Likewise for the second level. Let something similar hold of relation-words as well. We may say in general that words with the same grammatical function are to stand opposite one another. Each word occurring on the left has its determinate sense —at least we assume this—and likewise for each one occurring on the right. Now by means of this opposition the senses of the words on the left are also correlated with the senses of the words on the right. Let this correlation be one-to-one, so that on neither the left nor the right is the same thing expressed twice. We can now translate; not, however, from one language to another, whereby the same sense is retained; but into the very same language, whereby the sense is changed. We can now ask whether, in such a translation, there once more results from a proposition on the left a proposition on the right. Since a real proposition must express a thought, for this to be the case it will be necessary that to a thought on the left there again corresponds a thought on the right. If we make one of the spoken [natural] languages the basis of our considerations, this is of course doubtful. But let us here presuppose a logically perfect language. Then indeed, to every thought expressed on the left, there will correspond one expressed on the right. Even if one doubts this, at least it will be admitted that it may be so in some cases. Now

13. I have treated of concepts of the first and second level in my *Basic Laws of Arithmetic, 1*, (Jena, 1893), §§ 21 and 22. Here I confine myself to the following remarks. The subsumption of an object under a concept (Plato is a human being. Two is a prime number.) is familiar. Here we have only first-level concepts. First-level concepts can stand to second-level concepts in a relation similar to the one in which objects stand to first-level concepts. But here we must not think of subordination, because in the latter case, both concepts are of the same level. Consider the proposition "There is a prime number." Here we see the first-level concept *prime number* stand to a second-level concept in a relation similar to that in which 2 stands to the concept *prime number*.

let the premises of an inference be expressed on the left. We then ask whether the thoughts corresponding to them on the right are the premises of an inference of the same kind; and whether the proposition corresponding to the conclusion-proposition on the left is the appropriate conclusion-proposition of the inference on the right. In any case, the thoughts expressed on the right must be true in order to be premises. Let us assume that this condition is met.

One may now be tempted to answer our question in the affirmative, hereby appealing to the formal nature of the laws of logic according to which, as far as logic itself is concerned, each object is as good as any other, and each concept /428/ of the first level as good as any other and can be replaced by it; etc. But this would be overly hasty, for logic is not as unrestrictedly formal as is here presupposed. If it were, then it would be without content. Just as the concept *point* belongs to geometry, so logic, too, has its own concepts and relations; and it is only in virtue of this that it can have a content. Toward what is thus proper to it, its relation is not at all formal. No science is completely formal; but even gravitational mechanics is formal to a certain degree, insofar as optical and chemical properties are all the same to it. To be sure, so far as it is concerned, bodies with different masses are not mutually replaceable; but in gravitational mechanics the difference of bodies with respect to their chemical properties does not constitute a hindrance to their mutual replacement. To logic, for example, there belong the following: negation, identity, subsumption, subordination of concepts. And here logic brooks no replacement. It is true that in an inference we can replace Charlemagne by Sahara, and the concept *king* by the concept *desert,* insofar as this does not alter the truth of the premises. But one may not thus replace the relation of identity by the lying of a point in a plane. Because for identity there hold certain logical laws which as such need not be numbered among the premises, and to these nothing would correspond on the other side. Consequently a lacuna might arise at that place in the proof. One can express it metaphorically like this: About what is foreign to it, logic knows only what occurs in the premises; about what is proper to it, it knows all. Therfore in order to be sure that in our translation, to a correct inference on the left there again corresponds a correct inference on the right, we must make certain that in the vocabulary to words and expressions that might occur on the left and whose references belong to logic, identical ones are opposed on the right. Let us assume that the vocabulary meets this condition. Then not only will a conclusion again correspond to a conclusion, but also a whole inference-chain to an inference-chain, i. e. to a proof on the left there will correspond a proof on the right— always presupposing that the initial propositions on the right hold just as do those on the left.

Let us now consider whether a thought G is dependent upon a group of thoughts Ω. We can give a negative answer to this question if, according to our vocabulary, to the thoughts of group Ω there corresponds a group of true thoughts Ω', while to the thought G there corresponds a false thought G'. For if G were dependent upon Ω, then, since the thoughts of Ω' are true, G' would also have to be dependent upon Ω' and consequently G' would be true.

/429/ With this we have an indication of the way in which it may be possible to prove independence of a real axiom from other [real] axioms. Of course, we are far from having a more precise execution of this. In particular, we will find that this final basic law which I have attempted to elucidate by means of the above-metioned vocabulary still needs more precise formulation, and that to give this will not be easy. Furthermore, it will have to be determined what counts as a logical inference and what is proper to logic.

If, following the suggestions given above, one then wanted to apply this to the axioms of geometry, one would still need propositions that state, for example, that the concept *point,* the relation of a point's lying in a plane, etc. do not belong to logic. These propositions will probably have to be taken as axiomatic. Of course, such axioms are of a very special kind and cannot otherwise be used in geometry. But we are here in unexplored territory.

One can easily see that these questions cannot be settled briefly; and therefore I shall not attempt to carry this investigation any further here.

To whomever might wish to answer my expositions, I should like to recommend strongly that he begin by stating as clearly as possible what he calls an axiom, when he calls an axiom independent of others, and how he delimits the reference of the word "axiom" from that of the word "definition." Of course, if one asks only for the sake of asking, then, as Mr. Korselt fears, one can go on asking forever. But it would never occur to any reasonable individual to continue asking when there is simply no danger of misunderstanding. In scientific dispute, one must seek to find out as precisely as possible wherein the difference of opinion consists, so as to avoid a mere dispute over words. In the present case, one must decide whether an axiom is something audible, perhaps a sequence of words which grammar calls a proposition, and if such is the case, whether it is a pseudo-proposition or a real one; or whether perhaps an axiom is a thought, the sense of a real proposition. As long as the word "axiom" was used as a heading only, a fluctuation in its reference could be tolerated. Now, however, since the question of whether an axiom is independent of others has been raised, the word "axiom" has been introduced into the text itself and something is asserted or proved about what it is supposed to designate. It is now necessary to bring about complete agreement on its reference. In the first edition of his essay, Mr. Hilbert should already have given a definition of an /430/ axiom; or, if that seemed impossible, an explication. But perhaps at that time he had not yet become aware of his departure from the Euclidean usage of the word. But then he ought to have filled this gap in the second edition. As it stands, we remain completely in the dark as to what he really believes he has proved and which logical and extralogical laws and expedients he needs for this. It cannot even be recognized for certain whether Mr. Korselt's use of the word "axiom" agrees with Hilbert's. Therefore when I press for clarity, this is not to be shrugged off as useless questioning. If someone wanted to continue this dispute further without first making a serious effort to answer the questions just posed, he would only be making hot air—an enterprise in which I do not care to participate.

Part II

Frege and Thomae on Formal Theories of Arithmetic

Thoughtless Thinkers: A Holiday Chat

BY J. THOMAE IN JENA

> No matter how absurd and exaggerated
> it may seem: still, the human mind
> does not discover its laws in nature
> but rather imposes them upon it.
>
> Kant

Anyone wanting to base an arithmetic on a formal number-theory—a theory that does not ask what numbers are and what purpose they have, but instead asks what we need of numbers in arithmetic—will welcome the possibility of pointing to another example of a purely formal creation of the human mind. I believed that I had found such an example in the game of chess. Chess pieces are signs that have no content in the game other than the one assigned to them by the rules of the game. The locution that the signs are empty may lead to misunderstandings wherever an honest desire to understand is absent. I therefore believed that in the computing game, I could consider the numbers of arthimetic as signs that have no content in the game other /435/ than the one assigned to them by the rules of computation or the rules of the game. The system of signs of this computing game is constructed in the familiar way from the signs

$$0\ 1\ 2\ 3\ 4\ 5\ 6\ 7\ 8\ 9.$$

There is probably no language that is differentiated to such a degree that the sense of each of its words is completely unambiguous. So one says that the chess pieces have no significance; and yet it would surely be absurd to claim that the knight in chess has no significance (Bedeutung), or that a number has

From *Jahresbericht, 15* (1906), 434-38.

no significance (Bedeutung) in formal arithmetic—for example that it is not a relation of magnitude; and surely it would be absurd to claim that the number *e* has no significance (Bedeutung) in arithmetic, in the computing game. In chess, the content of a chess piece, for example that of the white queen's knight or of the black pawn of file *d,* is determined by its behavior vis-à-vis the rules of the game. Here the ambiguity of the word "behavior" once more manifests itself unpleasantly. In an ideal state all citizens behave in the same way with respect to the laws of that state; in a class state they behave differently. Animals, houses, etc. are also subject to the laws of the state, but they behave toward them differently from the way in which people behave. And now, on p. 101 of his *Basic Laws of Arithmetic,* I am informed by my colleague Mr. Frege that to behave oneself in a certain way means approximately the same thing as to conduct oneself in a certain way. Accordingly, chess pieces cannot behave vis-à-vis the rules of the game at all; at best it is the players who may, for example, make a wrong move.

If it is said that in the computing game a quotient whose denominator is zero has no significance (Bedeutung), then the reason given for this is that in the game both the signs themselves as well as those produced from them by means of the computation rules must be unambiguous. However, since the proposition

$$a = a, a \neq a$$

can easily be deduced from the form just mentioned, a quotient having zero as its denominator is inadmissible in formal arithemetic. And in general the demand that usage be unambiguous places certain restrictions on the creation of new signs in the computing game. For example, if I wanted to say that the two equations

$$3x = 1, 3\,x = 2$$

are mutually incompatible, yet create the sign $x = *$, which does service for both expressions, then $* = 1/3$ will have to be $* = 2/3$. Thus the sign would not be unambiguous, and consequently would be inadmissible in arithmetic. /436/ I have of course always been of the opinion that "black king" in chess signifies (*bedeute*) just as much as "Sirius" does in astronomy (Frege, p. 103).

It seems that even the word "identical" is not unambiguous. On p. 140 of the essay cited above, Mr. Frege posits the statement "The Morning Star is the same as the Evening Star" and apparently declares that the two are identical. Well, does the Morning Star have the same mass as the Evening Star? No. Does it have the same internal energy (thermal, chemical, etc).? No. Does it have the same direction of motion? No. The same conditions of illumination? No. But surely there will be some residue that is the same in both; consequently they are identical.—But when one says that the Morning Star is the same star as the Evening Star, a third element enters. The human mind, in its sovereign power of creation, imposing laws upon nature, forms the concept "star" and arranges this concept in such a way that the Morning Star is the same star as the Evening Star. Similarly, someone learning how to count abstracts from the differences of the counting blocks which he uses to learn this and equates them (*setzt gleich*) with one another. I believed that I had discerned the fruitfulness of the equality-sign in this possibility or capability of the human mind to abstract from the differences of certain things and to equate them with one another (*Elementary Function Theory*, p. 2), but I am thoroughly rebuffed by Frege.

Even with respect to chess have I been thoroughly mistaken, as Mr. Frege enlightens me. According to him, there are no theorems and proofs in this game. Therefore it would appear that the statement that castle and king together can put an opponent's king into checkmate on any of the side squares is a theorem only for me whose mind is insufficiently trained in logic; its proof is only an apparent one. Mr. Frege informs me that chess players do indeed think, but that they have no thoughts: they are thoughtless thinkers. This concept is probably distinct from that of a dreamer—at least I have never heard chess players being called dreamers. When I say that this bun is spread thick with butter, then this proposition, perhaps, expresses a thought, for it refers to something concrete, namely something edible. But when I say that the peculiarity of the knight becomes strikingly evident in the end-game of a stamma (smothered mate), then, if I understand Mr. Frege correctly, this statement apparently does not express a thought. We like to compare a chess player to a battle commander who devises a battle plan which he then changes during the course of battle, depending on the circumstances. I don't dare decide /437/ whether or not a battle commander also falls under the concept of a thoughtless thinker. Chess players, of course, think that they find deep, beautiful, piquant, and the like thoughts in their games and problems, so that even the word "thought" does not seem to be unambiguous.

With this, my chat is really at an end; but permit me to add a few remarks. — At one point in his essay, Mr. Frege expresses his indignation in these words: "Now Thomae lets even numbers grow." In the case in question I was of course thinking, in the manner of Till Eulenspiegel,* that we could let the number three grow, as in the following diagram:

$$3 \; 3 \; 3 \; . \; . \; 3 \; 3 \; . \; .$$

But there are those fateful little dots. These, at any rate, signify four more threes (cf. Frege, p. 135). Therefore, I can indeed let the number three grow, but only about eight times, and

* A prankster who played practical jokes on people by taking their words quite literally. [Trans.]

in any case only in a limited way. I learned in school that the number 1/3 cannot be represented by a decimal fraction. Instead, one wrote

$$1/3 = 0.33333..$$

But since the two little dots at any rate signify two more threes, 1/3 can be represented by a decimal fraction after all—by one with seven places.

It is a pity that Mr. Frege imposes such restraint upon himself when talking about the newer kinds of computations (species). He *appears* to descry the latter in the "approaches to a limit." But he adduces no example in which an approach to a limit is made by means other than those using the four familiar species that are extended by means of number-sequences or fundamental series so as to include the irrational numbers. More recent proponents of formal arithmetic have attempted to extend the computing rules formally to cover irrational numbers in that so far as common numbers are concerned, they consider this question to have been settled by Hankel. This explains the fact that many things which would be necessary for a complete development of their ideas are missing from their accounts. For example, in my own case, I am sorry to say, the precept which Hankel designates by "modulus of a mode of computation" is missing.

But my friend and colleague Frege does solve one problem in a fortunate manner. On p. 129 he says, "We may define thus: A row of houses is called an infinite series if none of the houses is the last one and if, following a rule yet to be given, we can keep on building houses." Since my esteemed colleague surely knows the location of a building site for such a street, since /438/ otherwise the concept "infinite" would resemble a snake bitting its own tail, he need only state it and the agrarian problem will be solved once and for all.

And after Frege has "once and for all done away with formal

number-theory," and after he has recognized that even his own attempt to give a logical foundation to numbers has failed (p. 253, Epilogue), it results that we are left with no numbers at all and according to him must come to the sad conclusion that

Mathematics is the most obscure of all the sciences.

Written in the dog-days of the year 1906.

Reply to Mr. Thomae's Holiday Chat

BY GOTTLOB FREGE IN JENA

On pp. 80–153 of volume 2 of my *Basic Laws of Arithmetic*,[1] I gave a critical analysis of the best-known theories of irrational numbers. The task was anything but pleasant. After all, destruction is always less satisfying than construction and also finds less recognition; for at first it almost exclusively creates nothing but enemies. I made a serious effort to understand unfamiliar modes of expression and trains of thought, since I recognized that /587/ such a work could be valuable only on the condition that it be most thorough and that one wish to be fair to even the most unfamiliar of theories. It required great renunciation on my part to pursue paths which from the very start I recognized as blind alleys, merely in order to prove in particular cases that they are blind alleys, and to do this without even the hope of learning something valuable from this occupation with unfamiliar thoughts. Nevertheless, I undertook this task for the sake of science, since the desire for what is true can arise only after the worthlessness of all surrogates has been recognized. I should like to hope not merely for my own sake, but also for that of science, that as many as possible try to understand my writings, my trains of thought, with equal thoroughness and with the same desire to be fair to them.

From *Jahresbericht, 15* (1906), 586-90.
1. Jena, Pohle, 1903.

I am convinced that with my critique of Thomae's formal arithmetic, I have destroyed it once and for all, and this conviction is merely confirmed by Mr. Thomae's holiday chat. Even if Mr. Thomae were correct in everything he said there, enough would remain of my critique to bring his formal theory to ruin. And how does he proceed in this chat? He repeats his claims without even mentioning my counterarguments. My distinctions —e.g. between a figure and a sign, between a game and the theory of a game—he suppresses, thus obscuring once more what I had illuminated. He adduces my propositions in such a way that the reader cannot know how I meant them or what reasons I had for holding them.[2] The matter is very simple! I think that those readers who care to take the trouble to compare Thomae's holiday chat with my expositions on pp. 96–139 of the second volume of my *Basic Laws* will find that there I have already refuted everything which Mr. Thomae adduces against me, with the possible exception of one point which I shall consider in more detail. Since this can be done briefly, let me adduce something from my Table of Contents. There, on p. ix, we read,

§ 93. There are neither theorems nor proofs nor definitions in a number-game, though there are such in the theory of the game.

This shows that part of Thomae's exposition is off target. Sometimes one might almost think that Mr. Thomae did not even read what he was attacking.

I don't know what value one might see in such a polemic. I wonder if there is anyone at all besides Mr. Thomae himself and Mr. Korselt[3] who believes in Thomae's number-game? I

2. Mr. Thomae, it seems, did not understand my propositions.
3. Compare his essay "Concerning the Foundations of Mathematics" (*Jahresbericht, 14* (1905)). From this it is obvious that he believes in Thomae's number-game only within limits. He says (p. 382) that the rules of Thomae's game can only be considered as suggestions for the formulation of a truly formal theory. Well, I shall wait until Mr. Korselt has produced one. Then we shall see further.

would regret having sacrificed so much time and effort in this matter less if Mr. Thomae had said something about whether for example my surmise about addition and abstraction in the number-game is correct. I raised many such questions. An answer to them would at least /588/ further the development somewhat, and some small utility for science might arise from this. But Mr. Thomae surely knows why he is silent on this point. For his major mistake consists precisely in this, that he continually abandons his role as a formal arithmetician and imports many things from nonformal to formal arithmetic that fit into the latter like square plugs into round holes; and he does so without noticing that formal arithmetic is superfluous if it has to presuppose nonformal arithmetic. In fact, if we abstract from formal arithmetic everything it has derived from nonformal arithmetic, almost nothing remains except a few incredible assertions. Mr. Thomae calls numbers signs, where by the latter he obviously means spatial, material things; but he never acts accordingly, and is miffed when someone else does. On this point, compare what he says in his holiday chat about the growing of three. In his case, the disease seems to be incurable.

What Mr. Thomae says on p. 436 about the word "identical" may be very useful for a defense attorney in a muder trial who has to invalidate the prosecuting attorney's proof of identity. I have prepared a speech appropriate for just such an occasion but shall refrain from presenting it, since I am sure that the reader himself is capable of preparing far nicer speeches of this sort. However, Mr. Thomae has brought out in me not only the lawyer but the poet as well. After reading what he said about abstraction, I vented my feelings in this verse:

> Abstraction's might a boon is found
> While man does keep it tamed and bound;
> Awful its heav'nly powers become
> When that its stops and stays are gone.

Yes, Mr. Thomae's abstraction is surely dangerous. For example, consider the case where the human mind, sovereign in its creative power, prescribing laws to nature, forms the concept *white powder* and arranges this concept in such a way that sodium bicarbonate is the same white powder as arsenic. We would do well to be very careful about abstracting. But we should also not forget its beneficial effects!

We abstract from the difference between figure and sign, and immediately figures are signs and signs figures. We abstract from that by which the relation of a figure to its role in the game differs from the relation of a sign to its reference (*Bedeutung*), and right away the two coincide. We abstract from the difference between a sign and what is signified, and immediately they coalesce. We abstract from the difference between rule and theorem and lo! they are identical. We abstract from the difference between formal and nonformal arithmetic, between game and science, and no one can keep them apart any longer. All this is very important; especially as regards Mr. Thomae's formal arithmetic. If abstraction did not perform these miracles, it could not survive a second longer.

But let us stop joking and once again be serious! All of us probably agree that time designations belong to a predicate, and that an /589/ object may have a property at one time which it does not have at another. At one time a person may have no knowledge at all of the multiplication tables, and at another time he may know them. Still, he is the same person, and for this we don't need abstraction.

Now Mr. Thomae continues thus:

> Similarly, someone learning how to count abstracts from the differences, if any, of the counting blocks which he uses to learn this and equates them *(setzt gleich)* with one another. I believed that I had discerned the fruitfulness of the equality-sign in this possibility or capability of the human mind to abstract from the differences of certain things and to equate them with one another.

Really now! Again and again this superficiality and weakness of thought which does not know whether by the word "equal" it wants to designate identity or something else. On the basis of what he said before, it seems that Mr. Thomae wants to achieve identity by means of abstraction. Very well! If through this the counting blocks become identical, then we now have only one counting block; counting will not proceed beyond "one." Whoever cannot distinguish between things he is supposed to count, cannot count them either. Even someone learning how to count must distinguish between the counting blocks; for example on the basis of their different distances from the edge of the table. The counting blocks must differ, and this difference must also be recognized. If abstraction caused all differences to disappear, it would do away with the possibility of counting. On the other hand, if the word "equal" is not supposed to designate identity, then the objects that are the same will therefore differ with respect to some properties and will agree with respect to others. But to know this, we don't first have to abstract from their differences. Since the objects will have to differ anyway, it really matters little whether they differ a bit more or a bit less, as long as they are distinguishable at all. To see, either physically or mentally, is to distinguish. Thomae's abstraction is nondistinguishing or nonseeing; it is not a power of insight or of clarity but one of obscurantism and confusion. If by the word "equal" we do not mean identity but merely agreement in some respect or other, then either we must state in what respect sameness obtains—e.g. equal in color, equal in odor, etc.—or this must be evident from the context. The word "equal" in itself says nothing if it is not somehow explained more precisely. And if the human mind can equate any objects whatever, it is especially meaningless, and the meaning of equating will also remain obscure. But don't let us imagine that by means of abstracting and equating we can achieve greater agreement than obtains in reality. We do not thereby alter anything in the things themselves. The only thing we can thus achieve is error with respect to these things. If the equality-sign in arithmetic were to have as nebulous and ungraspable a reference *(Bedeutung)* as this, it would be useless. We find that thinkers of Mr. Thomae's stripe want the one thing as well as the other, which latter, however, is incompatible with the first. What do they want to achieve by abstracting? They want—well, what they really want is identity; for of what use is partial agreement, since that already obtains even without abstraction! /590/ But do they really want to become serious about this identity? No: that they certainly do not! Things are supposed to become identical by being equated—as if that mattered to the things themselves. But distinguishable: yes, that they must remain. And to use Mr. Thomae's phrase, the external sign of this weakly behavior is the word "equal."

But why must I always repeat the same arguments? Twenty-two years ago, in my *Foundations of Arithmetic,* §§ 34–48,[4] I presented at length what must be considered when dealing with this question; and in § 48 in particular I indicated the reason why one is tempted to derive numbers via abstraction. Is my whole effort, then, to have been in vain? Is my attempt to discuss this question as thoroughly and penetratingly and understandably as possible, weighing arguments and counterarguments—is all this effort simply to have been wasted? Is all this to have been written into the wind? At that time, twenty-two years ago and even afterwards, even someone like Weierstrass could utter such peculiar balderdash when talking about the present subject. But now, surely, it is time to think more carefully about the matter before writing about it. If someone can confute my arguments, let him do so; if not, let him spare me— but I don't want to become unparliamentary. It seems that there are people from whom logical arguments run off as water off a duck's back. And apparently there also are opinions which,

4. Breslau, Koebner, 1884.

although repeatedly confuted[5] and although no serious attempt is ever made to refute this confutation, nevertheless continue to maintain themselves as though nothing had happened. I regret that I know of no admissible means, be they parliamentary or literary, of shooing these opinions back into their haunts, so that they never again dare emerge into the light of day.

5. Cf. my essay "Concerning Mr. Schubert's Numbers" (Jena, Pohle, 1899).

Explanation

BY J. THOMAE IN JENA

Twenty-two years ago, in a conversation with Mr. Frege, the latter quite unequivocally gave me to understand that he considered me incapable of grasping his more erudite deductions. He now announces this *urbi et orbi*. We are told that I am a weakly thinker who does things only by halves, and Mr. Frege only regrets not having admissible means, be they parliamentary or literary, of shooing such opinions as mine back into their haunts, never again to emerge into the light of day.

Such an action is surely the result of the same conscientiousness that compels us to call a fool a fool so as to warn especially our academic youth about him. I suppose that we may also consider the refutation of my views concerning the ever so difficult concept of identity by comparing my words to the speech of an attorney in a murder trial, a purely factual one!

/591/ I have no intention of entering into a polemic with Mr. Frege for the following reasons:

Such a controversy would be useless for all parties concerned. It would be a hopeless task to try and convince Mr. Frege himself. As he himself says, he approaches these questions with the infallible certainty, reached beforehand, that the point of view of formal arithmetic is mistaken. I also question whether he would ever really take the trouble to try and do justice to my

From Jahresbericht, *15* (1906), 590-92.

opinion. If I remember correctly, in his book he asks several times, "What could Thomae mean?" At the time of its publication, we were on a friendly footing and met several times a week; but not once did he ask me what I meant. Did he deem the effort too great? I find it superfluous to say anything on account of Frege; also, I lack the time.

But the matter would also hold little interest for other mathematicians. One does his arithmetic by simply applying the rules about the use of numbers, and is satisfied with their consistency. Another happens to believe that he has moreover found the "thing in itself" of numbers and speculates about it. But the use of numbers or of the system of signs for numbers is one and the same in both cases, and each is convinced that in so using them he is doing the right thing. When proving the transcendental nature of the number e it is quite irrelevant which one of the analysts' different points of view one adopts. The mathematician's interest in such a controversy would therefore be small; it would concern only the logician.

Furthermore, in such a battle I should be at a disadvantage as opposed to Mr. Frege: I lack the weapon of infallibility. But in this connection let me point out parenthetically that my chat must be granted one small success. Previously, Frege had disposed of formal arithmetic once and for all—i.e. had done away with it objectively. Now, however, he is merely convinced that he has destroyed it—therefore he has done away with it merely subjectively.

A special reply to Mr. Frege's essay would have been unnecessary if he had merely proceeded factually; that is, if he had merely questioned my powers of intellect. But he also accuses me of suppressing facts. I have suppressed the fact that Mr. Frege splits chess into game and theory.*

When writing my humorous sketch, I had indeed forgotten the Fregean split, only noticing it when I was correcting the

* The German *spalten* contains a clear allusion to the expression *Haare spalten*: to split hairs. [*Trans.*]

proofs; but despite this, I did not then take occasion to add anything, so as not to take up too much space needed for valuable reports. The result of this is that I must now ask for some space for this matter.

In the game of billiards, there is the skill and there is the theory of the game. The theory is concerned with the laws of recoil of balls on elastic bands or from other balls; with the influence of rolling or rotation on these laws, and how one can impart a certain rotation by means of the cue; etc. A player is usually little concerned with the theory, but instead looks on the game as a skill which he has acquired through practice. Nevertheless, the theory exists.

Differently with chess: Here there is no theory, although there are some attempts to construct a mathematical theory. The latter are /592/ of no concern here. There is this word "theory": but what does it mean? How are we to understand the word "theoretical" when in a(n older) chess book we read, "So far as actual practice is concerned, the Allgaier gambit is at least adequate. So far as theory is concerned, the value of this attack has not yet been settled."

In practice, a game is a bound game. Either it is explicitly agreed beforehand how much time each player has for a move, or it is tacitly assumed as a rule of good conduct that no player take more than a reasonable amount of time to think. Otherwise every player could avoid defeat by taking such a long time to think about his move that his opponent loses all patience and abandons the game. Theory, however, deals with the unbounded, free game. If someone wants to study a sequence of moves theoretically, he will play in such a way that he takes as much time to think about each move as he pleases, even taking back a move as often as he likes in order to make an apparently better one instead. The fact that the theoretician passes judgments about a certain position obtained does not differentiate him from the practitioner; it is only that the latter will be mistaken still more frequently than the former. In the theory, proofs are given *ab oculos* in such a way that the game is played as far and as frequently as necessary until the proof of the asserted proposition is given. Every problem (end-game) can be expressed as a theorem that is to be proved by the game. An end-game, however, is a *position* which (incidentally) happens to express a thought.

So much for my suppression of facts.—I should have thought that accusing an opponent of stupidity would have sufficed; why add suppression of facts?

I wrote these lines after having perused only the first four pages of Frege's polemic. I ask the editor to grant me space for this reply, although I don't even know the remainder of that polemic. But as I shall under no circumstances reply to it, I can answer without having read it to the end.

I said that I should not reply to Frege's essay. Nevertheless, in case I should live long enough to see another edition of my *Elementary Function Theory,* I reserve the right of replying in the introduction—where I treat somewhat of the nature of numbers—to one or the other of Frege's objections.

Jena, 11 November, 1906

Renewed Proof of the Impossibility of Thomae's Formal Arithmetic

BY GOTTLOB FREGE IN JENA

Opposed to formal arithmetic, there stands nonformal arithmetic.* The two differ as follows: In nonformal arithmetic, numerals really are signs: mere tools of research, intended to designate numbers, where the latter are the nonsensible objects of the science. In formal arithmetic, it is the numerals themselves that are the numbers; they are not mere tools but rather the very objects of the investigation.

The basic idea of Thomae's formal arithmetic may be expressed briefly as follows:

Mr. Thomae compares formal arithmetic to chess. To the chess pieces whose use is governed by the rules of the game, there correspond certain spatial and visible figures, produced by writing or printing, which Mr. Thomae calls signs. With respect to their use, too, rules are to be laid down. These are the rules of the computing game. However, we immediately encounter a difficulty. The beginning chess player is initially introduced to chess board and chess pieces as the things on and with which the game is played. Let us call such things game-pieces. The first question, then, is this: What are the game-pieces of the computing game? Mr. Thomae writes,

---

From *Jahresbericht, 17* (1908), 52-55.
* The German is *inhaltliche Mathematik.* [*Trans.*]

"The system of signs of the computing-game is produced in the familiar way from

$$0\ 1\ 2\ 3\ 4\ 5\ 6\ 7\ 8\ 9."[1]$$

Had he merely said that these numerals are the game-pieces of the computing game, we should have been satisfied. But now he apparently wants to say that the game-pieces are first produced from these numerals, and in the familiar way at that. How could this matter be familiar to us, since we still want to learn the computing game? Mr. Thomae here makes the mistake he commits time and again: He assumes as known that for which he still wants to lay the groundwork. For we certainly cannot know whether he counts "23," "2/3," "2-3," "2:(5 +3)," "$\sqrt{3}$," "3 > 2," "2 × 2 = 4,"

$$\text{"}\int_0^1 \frac{d\,\xi}{1+\xi}\text{"},$$

"2 × a," "1-(1/3) + (1/5)-(1/7)" ...[2] among the signs about which he is here talking, and which therefore are comparable to chess pieces. Anyone familiar with arithmetic certainly knows that such groups do occur. To that extent we can say that they were constructed out of these numerals in the familiar way. But surely, given this we still do not know whether Mr. Thomae considers them game-pieces. For aside from the numerals, other figures also occur: fraction-stroke, division-sign, root-sign, equality-sign, brackets, dots, letters, etc. Are these figures, then, not to occur in the computing game at all? Or are they, too, supposed to be game-pieces? Or is it that only certain groups in which they occur together with numerals are to be comparable to chess pieces? We don't know. And yet Mr. Thomae says, "in

1. This journal, *15*, 435.
2. I here use quotation-marks because I mean not what is designated by these groups in nonformal arithmetic, but rather these figures themselves as visible, corporeal signs.

the familiar way"! This uncertainty in which we are left is the mental fog that evidently is quite conducive to the success of the computing game. For now, let us console ourselves with the thought that, after all, we shall get to know the rules of the computing game. And of course given these, we shall presumably see with what sorts of things the latter are concerned.

The second question concerns the actions involved in playing the game. We are all familiar with the kind of action involved in playing chess. A piece is transferred from one square of the chessboard to another, or is completely removed from the board, etc. The rules of chess relate to these actions. The question as to the nature of the actions in the computing game is more difficult to answer. Perhaps they consist in this, that by means of writing, certain figures are formed or those already written are again erased. But how in the world can this be in doubt? After all, this must be evident from the rules of the computing game!

The first of the formulae which according to Mr. Thomae are supposed to contain the rules of this game, looks like this:

$$a + a' = a' + a$$

This formula, when understood in the sense of nonformal arithmetic, belongs to the latter as one of its theorems; it does not, however, belong to formal arithmetic. But it must not be understood that way, since nonformal arithmetic must not be presupposed. In formal arithmetic, however, nowhere previously has this formula been assigned a sense. It has been assigned a sense neither as a whole nor by separately explaining each of its parts. The vertical cross has not yet occurred at all. Therefore so far as formal arithmetic is concerned, this formula is quite senseless; as senseless as a configuration of chess pieces before it has somehow been accorded a sense or before any rule of chess has even been laid down. Now the rules of the computing game are of course supposed to give something like a content to the game-pieces—to the "signs". Whether this is possible or not is irrelevant: in neither case is this content /54/ already present before a rule has been laid down. Rules either command, forbid, or permit. But nowhere in formal arithmetic has a sign been explained by commanding, forbidding, or permitting something with respect to it. Therefore, already the first rule of the computing game vanishes into thin air, and with it the computing game itself.

One might perhaps attempt to avoid this conclusion by construing not the whole formula but merely parts of it—e.g. the equality-sign—in the sense of nonformal arithmetic. But exactly what sense is this? The equality-sign occurs only between signs or groups of signs; and here we must distinguish between two cases. In the first, as for example in

$$2/3 + 3/5 = 19/15,$$

a simple or a complex sign designating something stands on the left; similarly on the right. What a sign designates, I call its reference. Now the equality-sign designates a certain relation and is our means of expressing the fact that this relation obtains between the reference of the sign on the left and that of the sign on the right.

In the other case, as in

$$a + b = b + a,$$

we have letters. Of these we cannot say that they designate something, as do the numbers '2' and '1/2'. But although they have no reference, nevertheless they do contribute something to the sense of the proposition. I use the locution: They indicate so as to lend generality to the content of the proposition. By this I mean to say that no matter what numbers we understand by the letters,* we always get something true. Therefore, if our formula is to have a sense in non-formal arithmetic, we must

---

* The German has the singular; the plural rendering is in keeping with the whole context. [*Trans.*]

once more return to the first case when replacing the letters by numerals.

However, if in combination with the equality-sign there occur written or printed figures that neither have a sense in nonformal arithmetic nor are letters—as is the case with

$$\S\S = \pounds$$

then the sense which the equality-sign otherwise has in nonformal arithmetic cannot come into play. Such a combination has no sense.

A case that must be classed with the preceding is this: Where all the signs that occur can otherwise be understood in the sense of nonformal arithmetic but here are not to be understood in this way; as when in

$$2/3 + 3/5 = 3/5 + 2/3$$

'2/3' and '3/5' are not to be understood as numerals in the sense of nonformal arithmetic, but perhaps as game-pieces in the computing game.

/55/ Therefore, the equality-sign can be understood in the sense of nonformal arithmetic only when the whole formula of which it is a part is to be understood in the sense of nonformal arithmetic. Hence either the formula

$$a + a' = a' + a$$

is a theorem of nonformal arithmetic, or it is quite senseless. In neither case do we have a rule.

Mr. Thomae falls into error because he chooses as his game-pieces figures that look like the numerals of nonformal arithmetic. As far as the game itself is concerned, it should really be quite irrelevant what the game-pieces look like, as long as those that are distinguished in the rules are in fact clearly distinct, and as long as the actions of the game remain possible. In chess, for example, instead of castles, knights, bishops, queens, and kings, we could use pieces representing cannons, lancers, lieutenants, colonels, and generals, and could play with these just as well as with the traditional pieces. Similarly, it should be possible to play the computing game with figures that look quite different from the signs of arithmetic; e.g. these:

ꓤ ꓛ > < × § ∩ ∪ ∩̲ ∪̲ ∂℮ £ $³

But that doesn't work. Why doesn't it work? Why must the game-pieces agree with the signs of nonformal arithmetic? Because formal arithmetic cannot do without the sense which its objects have in nonformal arithmetic. It is comparable to a creeper twining around nonformal arithmetic, losing all hold once its support and source of sustenance are removed. Accordingly, formal arithmetic presupposes the nonformal one; its pretension of replacing the latter herewith falls to the ground.

Therefore, so far as the computing game is concerned, we have established the following:

1. We are not fully told with which game-pieces we are dealing.

2. We are left completely in the dark as to wherein the actions of this game consist.

We easily ought to be able to get clear on these two points if we were told the rules of the game; but

3. What we are offered as the rules of the game does not remove the above doubts. In formal arithmetic these formulae are senseless. In order to give them a sense, we should have to adduce nonformal arithmetic as extended to cover negative, fractional, etc. numbers—which in the first place is inadmissible, and in the second would not yield any rules.

3. Of course it would not occur to anyone to see a rule of the game in
< § ∂ × ∂ § <.

Concluding Remarks

BY GOTTLOB FREGE

In the preceding essay, I have combated a theory objectively and seriously. If Mr. Thomae knows something that can be opposed to it, then it is his duty to present it. There is no valid reason for keeping it back, except perhaps for continuing weakness. If, when seriously attacked, a doctrine is no longer defended, then by all general principles of scientific enterprise it must be considered refuted.

From *Jahresbericht, 17* (1908), 56.

Part III

Frege
On Formal Theories of Arithmetic

On Formal Theories of Arithmetic

BY GOTTLOB FREGE

I here want to consider two views, both of which bear the name
"formal theory." I shall agree with the first; the second I shall
attempt to confute. The first has it that all arithmetic propositions can be derived from definitions alone using purely logical means, and consequently that they also must be derived
in this way. Herewith arithmetic is placed in direct contrast
with geometry, which, as surely no mathematician will doubt,
requires certain axioms peculiar to it where the contrary of these
axioms—considered from a purely logical point of view—is just
as possible, i.e. is without contradiction. Of all the reasons that
speak in favor of this view, I here want to adduce only one
based on the extensive applicability of mathematical doctrines.
As a matter of fact, we can count just about everything that can
be an object of thought: the ideal as well as the real, concepts
as well as objects, temporal as well as spatial entities, events
as well as bodies, methods as well as theorems; even numbers
can themselves again be counted. What is required is really
no more than a certain sharpness of delimitation, a certain
logical completeness.* From this we may undoubtedly /95/

From *Sitzungsberichte der jenaischen Gesellschaft für Medizin und
Naturwissenschaft, 19* (1885), Suppl. 2, 94-104.
* *Vollkommenheit.* The pun on 'perfection' and 'completeness' which the
use of the German carries is typical for Frege. [*Trans.*]

gather at least this much, that the basic propositions on which arithmetic is based cannot apply merely to a limited area whose peculiarities they express in the way in which the axioms of geometry express the peculiarities of what is spatial; rather, these basic propositions must extend to everything that can be thought. And surely we are justified in ascribing such extremely general propositions to logic.

I shall now deduce several conclusions from this logical or formal nature of arithmetic.

First, no sharp boundary can be drawn between logic and arithmetic. Considered from a scientific point of view, both together constitute a unified science. If we were to allot the most general basic propositions and perhaps also their most immediate consequences to logic while we assigned their further development to arithmetic, then this would be like separating a distinct science of axioms from that of geometry. Of course, the division of the entire field of knowledge into the various sciences is determined not merely by theoretical but also by pragmatic considerations; and by the preceding I do not mean to say anything against a certain pragmatic division: Only it must not become a schism, as is presently the case to the detriment of all sides concerned. If this formal theory is correct, then logic cannot be as barren as it may appear upon superficial examination—an appearance for which logicians themselves must be assigned part of the blame. And the negative attitude of many mathemeticians toward anything philosophical is without any objective justification—at least so far as this attitude carries over into logic. This science is capable of no less precision than mathematics itself. On the other hand, we may say to the logicians that they cannot come to know their own discipline thoroughly unless they concern themselves more with mathematics.

My second conclusion is this, that there is no such thing as a peculiarly arithmetic mode of inference that cannot be reduced to the general inference-modes of logic. If such a reduction were not possible for a given mode of inference, the question would immediately arise, what conceptual basis we have for taking it to be correct. In the case of arithmetic, it cannot be spatial intuition, because thereby the discipline would be reduced to geometry—at least so far as some of its propositions are concerned. Nor, likewise, can it be physical observation, because thereby it would also be deprived of its general applicability, which extends far beyond the physical. /96/ We therefore have no choice but to acknowledge the purely logical nature of arithmetic modes of inference. Together with this admission, there arises the task of bringing this nature to light wherever it cannot be recognized immediately, which is quite frequently the case in the writings of mathematicians. I have done this in relation to the inference from n to $(n+1)$.

And as my second conclusion is concerned with modes of inference, so my third conclusion is concerned with definitions. In the case of any definition whatever we must presuppose as known something by means of which we explain what we want understood by this name or sign. We cannot very well define an angle without presupposing knowledge of what constitutes a straight line. To be sure, that on which we base our definitions may itself have been defined previously; however, when we retrace our steps further, we shall always come upon something which, being a simple, is indefinable, and must be admitted to be incapable of further analysis. And the properties belonging to these ultimate building blocks of a discipline contain, as it were *in nuce,* its whole contents. In geometry, these properties are expressed in the axioms insofar as they are independent of one another. Now it is clear that the boundaries of a discipline are determined by the nature of its ultimate building blocks. If, as in the case of geometry, these ultimately are spatial configurations, then the science too will be restricted to what is spatial. Therefore if arithmetic is to be independent of all particular properties of things, this must also hold true of its building blocks: they must be of a purely logical nature. From this there follows the requirement that everything arithmetical be reducible to logic by means of definitions. So, for example, I have replaced the expression "set", which is frequently used by mathematicians, with the expression customary in logic: "concept". Nor is this merely an irrelevant change in terminology, but rather is important so far as an understanding of the true state of affairs is concerned. The word "set" easily evokes the thought of a heap of things in space, as is evident for example from the expression "set of dishes";* and thus, like J. S. Mill, one very easily retains the childlike conception of a number itself as a heap or aggregate, or at least as a property of a heap, and in concert with K. Fischer takes calculating to be aggregative thinking. Hereby one completely forgets that one can also count events, methods, and concepts, where certainly we cannot make a heap of any of these. I characterize as a concept that which has number, and in so doing /97/ indicate that the totality, which is here our primary concern, is held together by characteristics, not spatial proximity, which latter can obtain only in special cases as a by-product of these characteristics, but which generally speaking is unimportant. Thus even the number zero, which otherwise is completely without bearer, becomes intelligible; for where is there a heap in which we could discover this number? But this merely as an example of the way in which what is arithmetical can be reduced to what is logical. Only in this way is it possible to fulfill the first requirement of basing all modes of inference that appear to be peculiar to arithmetic on the general laws of logic.

I now turn to the second of the two views that may be called formal theories; and herewith I come to the central part of my paper. This view has it that the signs of the numbers 1/2, 1/3, of the number II, etc. are empty signs. This cannot very well be extended to cover whole numbers, since in arithmetic we

---

* I substitute an English idiom for the German, since the latter, as is only too often the case, does not fit English usage very well. [*Trans.*]

cannot do without the content of the signs 1, 2, etc., and because otherwise no equation would have a sense which we could state —in which case we should have neither arithmetic truths nor an arithmetic science. It is curious that it is precisely the lack of its consequential application that has made the continued existence of this opinion possible. I therefore want to show that no one ever really puts this theory into practice or indeed could put it into practice, since if he did, it would very quickly become useless. Despite the emphatic assertion that the signs are empty and that it is they themselves that are the numbers, in the background there always hovers the thought that they do signify something and that it is these contents of signs that really are the numbers. This is indicated by the use of the word "sign"; for is something that does not designate anything and does not even have the purpose of designating anything a sign? In such a case I shall use the word "figure," so as to avoid confusing anyone by means of a wrong expression. By the way, it is un- necessary to lay any stress on the word "empty." The essen- tial point of this theory is that a number is called a sign, *ergo* a written sign; and it is really irrelevant if, over and above this, such a figure also serves as sign of a content, as long as it is not this content but the sign itself that is considered the number, i.e. that with which arithmetic is concerned. Let us for once take seriously the contention that 1/2 does not designate any- thing. Well, then it is an artifact consisting, perhaps, of printer's ink. The properties of this thing are geometrical, physical, and chemical ones. We now have to distinguish between /98/ halves that are printed, written with ink, or with pencil or chalk, where these differ in most of their properties. Thus, instead of the determinate singular object—the number 1/2—we now have a whole species of artifacts that manifest a certain relationship only with respect to their shape. And what great things could pos- sibly be inferred from this shape? Moreover, where is that prop- erty upon which everything here depends: that of yielding 1 when added to itself? Nothing can be seen of it. Whence, then, does

this property come? It is said that it is established by means of
a definition. Now the purpose of a definition is surely to indicate
what sense one connects with a word or sign. What is it, then,
that is to be defined here? The most obvious answer, and probably also what is most frequently meant, is that it is 1/2 that is to
be explained. But such a definition would not be in agreement
with the postulate that the figure 1/2 is devoid of content. Surely
it is impossible to give an explanation of that to which the sign
1/2 is supposed to refer and at the same time deny that the sign
has any content at all. Here we once again see that superficiality,
which does not quite put the theory of the emptiness of the sign
into practice. But as long as what is taken to be the number is
not the content but the sign itself, matters would not be improved
even if the sign were in fact granted a content. Since it is arbitrary what reference one wants to give to a sign, it follows that
the content of the sign will have these or those properties, depending on the particular choice made. Therefore it in part
depends on my will, which properties the content of the sign
has. But still, these will always be properties of the content of
the sign, not of the sign itself; hence they will not be properties
of the number in the sense of this formal theory. Mathematicians
—so someone might easily say—are very peculiar people; instead of investigating the properties a thing really has, they don't
care about them one iota, but using so-called definitions, ascribe
all sorts of properties to a thing that have absolutely no connection with the thing itself, and then investigate these properties.
Someone could just as easily hit upon the idea of branding his
fellow-citizen a liar by the simple expedient of [using] a definition. It would then be very easy to prove the truth of his
charge. He would merely have to say, "That follows immediately
from my definition." Indeed; it would follow from the definition
just as rigorously as it follows from the definition "this chalk-
figure has the property of yielding 1 when added to itself" that
when added to itself, the figure yields 1. In this way we could
certainly define many things; /99/ a pity, that the things them-

selves would not care one iota, nor give up their old properties or assume new ones solely for the sake of our definition.

This state of affairs, which is really quite obvious, is sometimes obscured by the fact that it is not clearly recognizable precisely what is supposed to be defined: the sign 1/2, or the plus-sign, or the equality-sign, or perhaps a combination of several of these signs. But surely it is a justifiable demand that only one sign be explained in any one definition, and that it be clearly recognizable, which sign it is. This, too, is an error: to pass off as the definition itself what in fact is merely a rule for constructing definitions or a mere assertion to the effect that one is defining. I should be committing this error if I were to say, for example, "I define the plus-sign in such a way that $1/2 + 1/2 = 1$." This is just as if I were to say, "I shall now fly into the air," yet at the same time remained standing on the ground. The important thing is not that one say that one is defining; rather, it is that one actually do so, and in such a way that what is wanted is actually achieved.

Anyone ignorant of the historical development of the subject will scarcely understand how someone can come to represent empty signs as the proper objects of arithmetic investigations. Given what was said above, this may perhaps seem quite nonsensical as long as one is unaware of any motive for this claim which, after all, seems to create quite unnecessary difficulties. Therefore I shall have to delve a little deeper into the matter. Even in prescientific times, because of the needs of everyday life, positive whole numbers as well as fractional numbers had come to be recognized. Irrational as well as negative numbers were also accepted, albeit with some reluctance—the Greeks could not quite bring themselves to recognize them—and it was with even greater reluctance that complex numbers were finally introduced. The overcoming of this reluctance was facilitated by geometric interpretations; but with these, something foreign was introduced into arithmetic. Inevitably there arose the desire of once again extruding these geometrical aspects. It appeared contrary to all reason that purely arithmetic theorems should rest on geometric axioms; and it was inevitable that proofs which apparently established such a dependence should seem to obscure the true state of affairs. The task of deriving what was arithmetic by purely arithmetic means, i.e. purely logically, could not be put off.

A solution that immediately suggested itself was to define these higher numbers by means of their properties: /100/ for example to say, "$\sqrt{2}$ is something which, when multiplied by itself, yields 2." Such a definition really presupposes that multiplication has previously been defined in such a way that it does not necessarily apply to whole numbers; else one would easily revert to the previously criticized error of defining two things at once—$\sqrt{2}$ and multiplication. But this merely *en passant*. So far, by means of such a definition one has merely obtained a concept, and there arises the question whether this concept is empty or fulfilled. As long as it has not been proved that there exists one and exactly one thing of this kind, it would be a mistake in logic to immediately use the definite article and say, "the number that when multiplied by itself yields 2," or "the square root of 2." Until such a proof has been given, one may only use expressions such as "a square root of 2," "all square roots of 2," in which "square root of 2" is treated like a concept-word and therefore not as a term of an equation. So far, nothing having the desired property has been obtained by means of such a definition; for the concept of a number which when multiplied by itself yields 2 no more has the property of yielding 2 when multiplied by itself than the concept of a right-angled triangle is a triangle or has a right angle. Now it is especially important for the derivation of many theorems that there be such higher numbers. And just as for many proofs in geometry one needs points or lines which do not occur in the theorems themselves, and just as in each of these cases it is then necessary to show that there are such auxiliary points or lines; so too in arithmetic, many theorems are proved with the aid of $\sqrt{-1}$, where this magnitude does not itself occur in the theorems. Now if there were simply no number whose square is − 1, these proofs would collapse. It seems that such an existence-proof is now to be rendered superfluous by saying, "The figure which I am now writing is itself the number, is itself the object of our considerations." But this clearly makes things a little too easy. If it were correct, it would apparently always lie within our power to prove existence, and consequently we could prove the most wondrous things. It is for this reason that definitions are required to be noncontradictory—about which I shall say something later. At this juncture, it may suffice to point out that the properties indicated by a definition do not belong to the sign; and that if a sign is not supposed to have any reference, /101/ a definition of it will also be impossible. Thus the situation may surely be stated like this: Either a number is a written figure, in which case, of course, one cannot very well doubt its existence, although at the same time it also does not have the properties required of it; or a number is the content of a sign, in which case one will have to give a proof to the effect that the sign does indeed designate something and is not perhaps empty, contrary to all intentions. In the latter case, existence can no longer be established merely by pointing to the sign, and the latter lessens in significance to that of an unessential expedient which need not be mentioned at all in the initial justification of numbers.

Now one might perhaps attempt to avoid this difficulty by defining not numbers, but rather methods of computation. The following example will show why this does not work. The addition of fractions, for example, must be explained generally, perhaps like this: The sum of a/b and c/d is

$$\frac{a \times d + b \times c}{b \times d}.$$

It is quite impossible to prove in this way that the sum of $1/2$ and $1/2$ is 1. For what we obtain is

$$\frac{1 \times 2 + 1 \times 2}{2 \times 2}.$$

This is an empty figure and not the number 1. According to the above, one also could not say that 1/2 is equal to 3/6, for both are merely figures. The situation changes radically when one takes these figures to be signs of contents; in that case, the equation states that both signs have the very same content. But if no content is present, the equation has no sense.

However, one can give the matter another turn that seems to favor this formal theory. One could say something like this: "We don't define at all, but merely stipulate rules in accordance with which one can move from given equations to new ones, just as one can give rules for moving chess pieces. As long as an equation does not contain only positive numbers, it no more has a sense than the position of chess pieces expresses a truth. Now in virtue of these rules it may happen that an equation of positive whole numbers finally does occur. And if the rules are of such a kind that if one has started with true equations they can never lead to false conclusions, then only two cases are possible: either the final equation is senseless, or it has a /102/ content about which we can pass judgment. This last will always be the case if it contains only positive whole numbers; and then it must be true, for it cannot be false." The only question is, how the rules are to be arranged so that nothing false ever results. One might say something like this: "If these rules contain no contradictions among themselves and do not contradict the laws of positive whole numbers, then no matter how often they are applied, no contradiction can ever enter in. Consequently, if the final equation has any sense at all, it must also be noncontradictory, and hence be true." Now already this last is a mistake; for a proposition may very well be noncontradictory without for all that being true. Noncontradictoriness, therefore, does not suffice. And even if it did, still, it would first have to be proved. It seems that this is frequently deemed unnecessary;

but surely the example of indirect proofs shows that contradictions are not always evident but instead are often brought to light only by a series of inferences. It is impossible to say at the outset, how many inferences will be needed for this. Therefore if after a series of inferences no contradiction has been discovered, one still cannot know whether upon further continuation of the inference-chain a contradiction might not come to light after all. A proof of noncontradictoriness, then, cannot be given by saying that these rules have been proved as laws for the positive whole numbers and therefore must be without contradiction; for after all, they might conflict with the peculiar properties of the higher numbers, e.g. that of yielding -1 when squared. And in fact, not all rules can be retained in the case of complex higher numbers in a three-dimensional domain. At least the theorem that a product can be zero only if one of the factors is zero—a theorem fundamental to algebra—must be dropped. It is therefore evident that in virtue of the peculiar nature of the complex higher numbers, there may arise a contradiction where so far as the positive whole numbers are concerned, no contradiction obtains.

Therefore so far as this formal theory is concerned, the confidence that no contradiction is contained in the computation-rules stipulated for ordinary complex numbers no doubt rests on the fact that so far, none has been found. If one then proves the formula

$$cos^n \alpha = cos^n \alpha - \frac{n(n-1)}{1 \cdot 2} cos^{n-2} \alpha \; sin^2 \alpha + \ldots$$

with the aid of the complex numbers by raising $(cos\alpha + isin\alpha)$ /103/ to the n-th power, one is really moving in a circle. Despite this proof, the formula might still be false; and if by some other means one were to arrive at an equation that is incompatible with the former, one would have to say that it here becomes obvious that the postulated rules for complex numbers contain a contradiction. The proof of the above formula is based on the noncontradictoriness of the rules for complex numbers, and the latter in turn is established by the fact that no contradiction has ever been found. In any other discipline we should rest content with such a great probability; mathematics, however, wants to prove all of its theorems, and the preceding is not a proof—at least not so far as this formal theory is concerned.

Undoubtedly one moves in a similar circle when one says that numbers exist if one can compute with them. In the first instance, one would be quite justified in asking what inner connection there obtains between existence and computing. Then, for the answer to be of any use at all, one would have to have a definition of computing, so as to know whether what one is doing with a sign is to be called computing. Can one not also compute with diverging series? Judging by appearances, yes; only sometimes what results is false. This suggests the following definition: What is to be called computing is an operation that never yields false results. And with this we have once more landed in a circle. What means are there for proving noncontradictoriness? I see no other principle that could serve here except the following: that properties which are found on one and the same object can never stand in contradiction to one another. But if one had such an object, this formal theory would be superfluous. It therefore seems to me that it is unlikely that a strict proof of the noncontradictoriness of the computation-rules can succeed without going beyond the limits of this formal theory. But even if it should succeed, it would still not suffice; for not everything that is noncontradictory is true.

Clearly we are here faced with a difficulty consisting in the fact that we are defining objects, whereas otherwise we are merely concerned with defining concepts; 1/2, for example, must be considered an object, albeit one that is neither sensibly perceptible nor spatial. Despite this nonspatiality and nonreality, 1/2 is not a concept in the sense that objects /104/ can fall under it. One cannot say "this is a 1/2," as one can say "this is a right angle"; expressions like "all 1/2" or "some 1/2" are equally as inadmissible. Rather, 1/2 is treated as a single determinate object, as is evident from the expression "the number 1/2," and from the fact that it stands on one side of the equality-sign. It would here be going too far to indicate how I conceive of the solution of this difficulty. Let me merely say this, that the case of fractions, negative numbers, etc. is not essentially different from that of positive whole numbers; and that therefore a reduction of all propositions or equations to those that deal with positive whole numbers merely pushes the difficulty back one step further and does not solve it. The only reason why this is not noticed quite so easily is the fact that usually one does not feel the need to justify the most primitive of all numbers.

Index

Abstract, xxxiv, 117, 124–26
Abstracting, 123–26
Abstraction, xix, 123–25. *See also* Identity
Agreement, 125, 126
Ambiguity, ix, xxxvii, 66–68, 79, 116. *See also* Concept, second-level
Ambiguous, xxvi, xxvii; sign, 66–69, 73
Analysis, 143
Analytic, 61
Analyze, 57. *See also* Construct
Antecedent, xxxiv, xxxv, 70, 71, 106; clause, 52; proposition, xxxvii, 69, 70, 75, 77, 80, 89, 98; pseudo-proposition, xxxv–xxxviii, 72–74, 76–81, 85, 88, 89, 92, 100; real proposition, xxxvii
Applicability, 21, 143
Application, 14, 75, 81, 96
Apply, 14, 31, 41, 45
Apprehend, xxxiii. *See also* Intuition
Apprehensible, 63
Arbitrary stipulation. *See* Stipulation
Argument, xxi. *See also* Inference

Arithmetic, 53–54, 65, 70; formal, 116, 119, 122–24, 132, 134; formal vs. nonformal, 132–34, 141–53 passim; reducible to logic, 142–44, 148. *See also* System, formal
Article, definite, 148; indefinite, 4
Assert, xxxvii, 7, 35, 50–53
Assertion, 24, 38, 41, 44, 48, 61; basic, 40, 42, 76, 92
Assign, 62, 65, 68, 81. *See also* Reference
Assume, 8
Assumption, 13–14
Axiom, 6–39 passim, 76–81, 86, 98; defining, xxxviii, 9, 25, 52–53, 56; definition of, 111; vs. definition, 53–59; explication of, 111; expression, xxx; false, 104; first-level, xiii, xxxii, xxxix, xl; -group, 11, 26, 42; invalid, xxx, 28, 44, 53, 96; of parallels, 10, 11, 21, 37, 42, 54, 102; real, xxxviii, 102, 106; as rule, 95, 146; second-level, xiii; -structure, 13; as a thought, 25n, 103; true, 81; valid, 28. *See also* Independence, of axioms